The Editor's Guide to
DaVinci Resolve 18

T0389852

The Editor's Guide to DaVinci Resolve 18

Chris Roberts

© 2022 by Blackmagic Design Pty Ltd

Blackmagic Design

www.blackmagicdesign.com

To report errors, please send a note to learning@blackmagicdesign.com.

Series Editor: Patricia Montesion

Editor: Dan Foster

Contributing Authors: Arthur Ditner, Daria Fissoun, and Dion Scoppettuolo.

Cover Design: Blackmagic Design

Layout: Danielle Foster

ISBN 13: 979-8-9872671-2-7

Contents

Foreword

Welcome to The Editor's Guide to DaVinci Resolve 18.

DaVinci Resolve 18 is the only post-production solution that combines editing, color correction, visual effects, motion graphics, and audio post-production all in one software tool! Its elegant, modern interface is fast to learn for new users yet powerful enough for the most experienced professionals. DaVinci Resolve lets you work more efficiently because you don't have to learn multiple apps or switch software for different tasks. It's like having your own post-production studio in a single app!

DaVinci Resolve 18 adds Blackmagic Cloud support for remote collaboration, DaVinci proxy workflow, new Resolve FX, intuitive object masking, improved subtitling for editors, Fairlight fixed bus to FlexBus conversion, and so much more!

Best of all, Blackmagic Design offers a version of DaVinci Resolve 18 that is completely free! We've made sure that this version of DaVinci Resolve includes more features than any paid editing system. That's because at Blackmagic Design we believe everybody should have the tools to create professional, Hollywood-caliber content without having to spend thousands of dollars.

I invite you to download your copy of DaVinci Resolve 18 today and look forward to seeing the amazing work you produce!

Grant Petty
Blackmagic Design

Acknowledgments

We would like to thank the following individuals for their contributions of media used throughout the book:

— Brian J Terwilliger, Terwilliger Productions for Living In the Age of Airplanes.

— Nuyen Anh Nguyen, Second Tomorrow Studios for Hyperlight.

— Chris Lang, Aaron Walterscheid, Nathan LeFever, and Sherwin Lau for Organ Mountain Outfitters content. "Furever Glass" music composed and performed by Matt Carlin.

— Miss Rachel's Pantry in Philadelphia, PA.

— Miserable Girl—Jitterbug Riot, EditStock ad

— HaZ Dulull for Sync footage—Sync is a short proof-of-concept film written, produced, and directed by Hasraf "HaZ" Dulull and is the property of hazfilm.com.

— Citizen Chain footage.

About the Author

Chris Roberts has spent the last 25 years editing everything from online corporate promos to broadcast television, with editing credits that include the BAFTA Award-winning series The Great House Giveaway.

He has been delivering video editing training for nearly 20 years and has trained university students and staff; broadcast journalists; and sports, factual, and drama editors. As a Blackmagic Certified Master Trainer, he has been responsible for delivering DaVinci Resolve training to end users and other trainers around the world, both in person and remotely.

Over the years, he has also written articles on editing techniques and editing software for a variety of magazines and online publications, as well as writing several other books, including The Beginner's Guide to DaVinci Resolve 18.

Chris lives in Worcestershire, UK, with his partner Samantha and, when not working, enjoys reading post-apocalyptic fiction, listening to hard rock and blues music, and binge-watching the TV programs he has invariably missed.

This book is dedicated to the memory of his dearly loved and sadly missed mum, Maureen.

www.chrisroberts.info

Getting Started

Welcome to **The Editor's Guide to DaVinci Resolve 18**, the official Blackmagic Design Training and Certification book that teaches editors, artists, and students how to edit in DaVinci Resolve. All you need is a Mac or Windows computer, the free download version of DaVinci Resolve 18, and a passion to learn and tell your story!

This official step-by-step training guide covers the basics of editing video and audio content so you can start creating your own Hollywood-caliber film and video today!

What You'll Learn

— Apply advanced editing and trimming techniques for multiple genres as used by professional editors around the world.

— Generate and manage proxy media with the Blackmagic Proxy Generator.

— Perform variable speed changes to enhance action.

— Use trimming tricks and real-time dynamic trimming.

— Apply multicamera syncing and editing techniques.

— Organize large projects efficiently using metadata and smart bins.

— Build complex composites on the edit page.

— Use keyframes to create sophisticated animations.

— Edit and mix audio for stereo and surround sound.

— Create, import, and edit subtitles for different languages.

— Deliver projects for online distribution, broadcast TV, and streaming services.

— Discover dozens of tips and tricks throughout the book that will transform how you work!

The Blackmagic Design Training and Certification Program

Blackmagic Design publishes several training books that take your skills farther in DaVinci Resolve 18. They include:

— *The Beginner's Guide to DaVinci Resolve 18*

— *The Colorist Guide to DaVinci Resolve 18*

— *The Editor's Guide to DaVinci Resolve 18*

— *The Fairlight Audio Guide to DaVinci Resolve 18*

— *The Visual Effects Guide to DaVinci Resolve 18*

Whether you want an introductory guide to DaVinci Resolve or want to learn more advanced editing techniques, color grading, sound mixing, or visual effects, our certified training program includes a learning path for you.

After completing this book, you are encouraged to take a 1-hour, 50-question online proficiency exam to receive a Certificate of Completion from Blackmagic Design. The link to the online exam can be found on the Blackmagic Design training webpage. The webpage also provides additional information on our official Training and Certification Program. Please visit

www.blackmagicdesign.com/products/davinciresolve/training

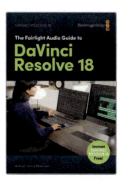
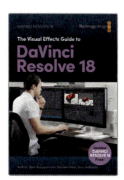

System Requirements

This book supports DaVinci Resolve 18 for macOS and Windows. If you have an older version of DaVinci Resolve, you must upgrade to the current version to follow along with the lessons.

> **NOTE** The exercises in this book refer to file and resource locations that will differ if you are using the version of software from the Apple Mac App Store. For the purposes of this training book, we recommend that macOS users download and use the DaVinci Resolve software from the Blackmagic Design website rather than from the Mac App store.

Download DaVinci Resolve

To download the free version of DaVinci Resolve 18 or later from the Blackmagic Design website:

1   Open a web browser on your Windows or macOS computer.

2   In the address field of your web browser, type: **www.blackmagicdesign.com/products/davinciresolve**.

3   On the DaVinci Resolve landing page, click the Download button.

4   On the download page, click the option appropriate to your computer's operating system.

5   Follow the installation instructions to complete the DaVinci Resolve installation.

When you have completed the software installation, follow the instructions in the following section to download the content for this book.

Copying the Lesson Files

The DaVinci Resolve lesson files must be downloaded to your macOS or Windows computer to perform the exercises in this book. After you save the files to your hard drive, extract the file and copy the folder to your Movies folder (macOS) or Videos folder (Windows).

To Download and Install the DaVinci Resolve Lesson Files

When you are ready to download the lesson files, follow these steps:

1   Open a web browser on your Windows or macOS computer.

2   In the address field of your web browser, type:
    www.blackmagicdesign.com/products/davinciresolve/training

3   Scroll the page until you locate the *The Editor's Guide to DaVinci Resolve 18.*

4   Click the Lesson Files link to download the media. The file is roughly 16.5 GB in size.

5   After downloading the zip file to your macOS or Windows computer, open your Downloads folder and double-click R18_Editors_Lessons.zip to unzip it if it doesn't unzip automatically. You'll end up with a folder named "R18 Editors Guide" that contains all the content for this book.

6   From your Downloads folder, drag the R18 Editors Guide folder to your Movies folder (macOS) or Videos folder (Windows). These folders can be found within your User folder on either platform.

You are now ready to begin Lesson 1.

Getting Certified

After completing this book, you are encouraged to take the 1-hour, 50-question online proficiency exam to receive a Certificate of Completion from Blackmagic Design. The link to this exam can be found on the Blackmagic Design training webpage:

https://www.blackmagicdesign.com/products/davinciresolve/training

Introducing Blackmagic Cloud

DaVinci Resolve is the world's only complete post-production solution that lets everyone work together on the same project at the same time. Traditionally, post-production follows a linear workflow with each artist handing off to the next, introducing errors and mountains of change logs to keep track of through each stage. With DaVinci Resolve's collaboration features, each artist can work on the same project, in their own dedicated page with the tools they need.

Now Blackmagic Cloud lets editors, colorists, VFX artists, animators, and sound engineers work together simultaneously from anywhere in the world. Plus, they can review each other's changes without spending countless hours reconforming the timeline.

Simply create a Blackmagic Cloud ID, log in to the online DaVinci Resolve Project Server, and follow the simple instructions to set up a new project library—all for one low monthly price!

Once created, you can access this library directly from the Cloud tab in the Project Manager to create as many projects as you need—all stored securely online. Then invite up to 10 other people to collaborate on a project with you. With a simple click, they can relink to local copies of the media files and start working on the project immediately, with all their changes automatically saved to the cloud.

Enabling Multiple User Collaboration for your project means that everyone can work on the same project at the same time—edit assistants, editors, colorists, dialogue editors, and visual effects artists can now all collaborate wherever they are in the world in a way never before possible.

Media Sync with Blackmagic Cloud Store

Now you don't need to buy expensive proprietary storage that needs an entire IT team to manage! Blackmagic Cloud Store has been designed for multiple users and can handle the huge media files used by Hollywood feature films. You can also have multiple Blackmagic Cloud Stores syncing the media files with your Dropbox account so that everyone has access to the media files for the project.

To find out more about these exciting workflows, visit **blackmagicdesign.com/products/davinciresolve/collaboration**

Building the Rough Cut

Editing is so central to cinematic storytelling that director Francis Ford Coppola once said, "The essence of cinema is editing." This book explores this "essence" of cinema, as applied to the art and craft of editing and storytelling, through the robust and powerful editing features found in DaVinci Resolve 18.

Whether you're working to cut the latest cinematic blockbuster, a fast-turnaround commercial spot, an episodic TV show, or an entire web series, the tools, technology, and functionality available to you in DaVinci Resolve, together with the techniques discussed and demonstrated throughout this book, will help you achieve your vision.

Time

This lesson takes approximately 60 minutes to complete.

Goals

The edit page supports the approach to nonlinear editing that has been battle-tested by editors around the world for decades. In this lesson, you'll learn the basics of those techniques as they apply to building a rough cut. In Lesson 2, you will then learn how to refine this rough cut into a final, polished masterpiece.

Setting Up a Project

Editing is often an iterative process that requires you to build a coherent story from disparate pieces of footage. While there are many generally accepted workflows for assembling these sounds and pictures, unfortunately there is no definitive "right way" to edit; every step the editor takes has its own unique aims, considerations, and consequences.

With that said, you'll start your exploration of editing in DaVinci Resolve by putting together a rough cut of a 1-minute social media promo for New Mexico outdoor clothing brand Organ Mountain Outfitters. Along the way, you'll gain an appreciation of some of the thought processes and happy accidents that often occur in editing suites around the world.

You will begin this lesson by importing a project that already contains the clips needed for the edit, and which has also been organized for you using a variety of techniques that you'll explore further in later lessons.

1   Open DaVinci Resolve to display the Project Manager.

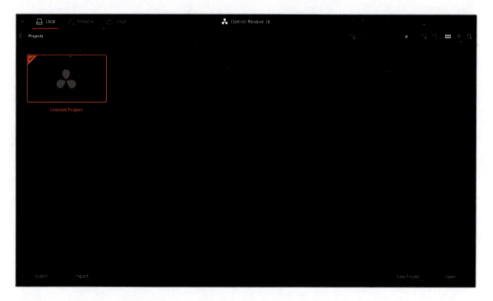

To keep the projects used in this book separate from any of your other current projects, you will create a new local project library on your system.

Project libraries are collections of projects that can reside locally on your computer or on directly attached storage media, on a network location that can be shared by people on the same network, or in the Blackmagic Cloud, where project libraries can be accessed online by anyone you choose to invite. Network and cloud-based project libraries also have the advantage that they can be used to collaborate, so that multiple DaVinci Resolve users can work on the same project simultaneously.

To set up a cloud-based project library, you will need a Blackmagic Cloud ID. You can register for a free account at **cloud.blackmagicdesign.com**. Once registered, you can use the project server to set up a project library in the cloud for a single, low monthly cost that you can cancel at any time. You can access any project libraries you have created in Blackmagic Cloud by clicking the Cloud option in DaVinci Resolve's Project Manager and logging in using your Blackmagic Cloud ID. Anyone hosting a project library in Blackmagic Cloud can invite up to 10 additional uses to collaborate in real time for no additional cost. Project libraries on a network can be set up and managed using the free DaVinci Resolve Project Server, available for download from the DaVinci Resolve support pages on the Blackmagic Design website at **https://www.blackmagicdesign.com/support/family/davinci-resolve-and-fusion**.

For the purposes of this book, you will work with a local project library.

2   Ensure that Local is selected at the top of the Project Manager, and then click the Show/Hide Project Libraries button to reveal the project libraries.

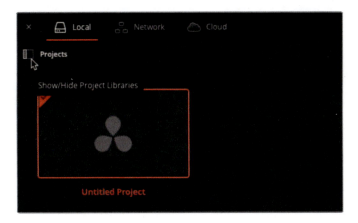

> **NOTE** If you haven't yet created any project libraries, you will see only the default local project library, called "Local Database." In versions of DaVinci Resolve prior to version 18, project libraries were called databases.

3   Click the Add Project Library button at the bottom of the project libraries list.

The Add Project Library window appears.

4   In the Add Project Library window, click the Name field, type **R18 Editors Guide**, and then click the Browse button.

You'll need to select (or create) an empty directory on your system for a new project library. For the purposes of the lessons in this book, you can use the folder provided in the downloaded R18 Editors Guide folder.

5 Navigate to R18 Editors Guide/Editors Guide Project Library and click Open to select this folder as the location for your new project library.

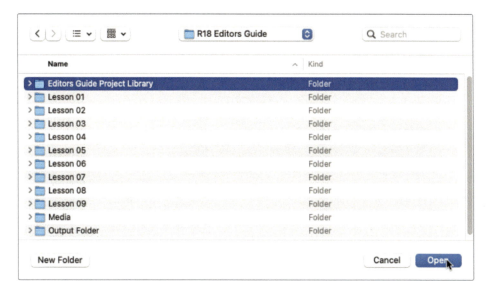

> **NOTE** If this is not your first time working through this exercise, and you've already created a project library in this folder, simply create a new folder in the R18 Editors Guide folder and choose that as the location for your new project library.

6 In the Add Project Library window, click Create to create your new project library.

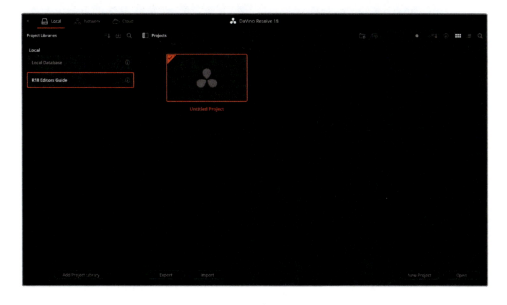

The new project library appears in the list of project libraries connected to DaVinci Resolve.

TIP To switch project libraries, simply click on the desired project library from the list to reveal the projects contained within that library. Projects can be copied from one project library to another using the Copy Project To button at the top of the Project Manager.

Having successfully created a new empty project library, you can import the project file needed for this first lesson.

7   Click the Import button.

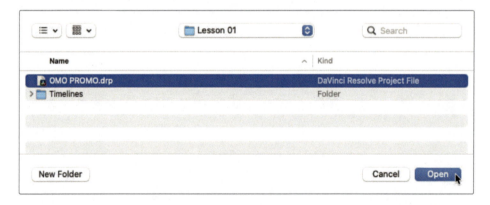

8   Navigate to R18 Editors Guide/Lesson 01 and select the file OMO Promo.drp.

Name	Kind
OMO PROMO.drp	DaVinci Resolve Project File
Timelines	Folder

New Folder   Cancel   Open

NOTE Files using the extension .drp are DaVinci Resolve Project files.

9   Click Open.

> **TIP** Alternative ways of importing a project include dragging .drp files directly onto the Project Manager, right-clicking in an empty area of the Project Manager and choosing Import Project, or choosing File > Import Project.

The project is imported into the project library and appears in the Project Manager.

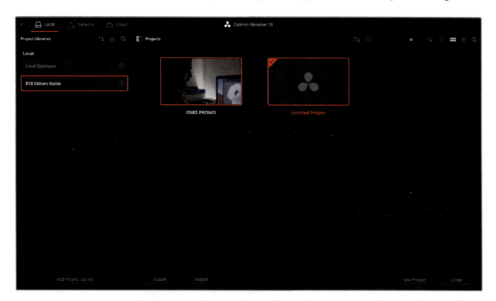

DaVinci Resolve does not "open" project files directly. Instead, imported .drp files are copied into the current project library. It is this imported project that you will be able to open and make changes to, not the original .drp file you selected for import. Clicking the Export button allows you to export a selected project(s) from the project library to individual .drp files, but any further changes to the project can only be made to the project in the project library. You can find out more information about managing project libraries in *The Beginner's Guide to DaVinci Resolve 18* (blackmagicdesign.com/products/davinciresolve/training).

10 Double-click the OMO Promo project to open it.

The project opens in DaVinci Resolve on the page last used in the application.

11 If necessary, click the edit page button.

12 Select Workspace > Reset UI Layout to reset the edit page workspace to the default configuration.

You have successfully imported and opened a project but, before you begin editing with the clips in this project, you'll need to relink the clips in this project to their media files on your computer's hard drive.

Relinking the Media Files

The reason why clips may go offline is often because the associate media files have been moved or renamed on the original storage drive. To prevent this from happening, once you have imported the clips you should leave them in their original locations on your computer. If you need to move them for any reason, you can use DaVinci Resolve's Media Management feature. You'll learn more about media management in Lesson 9, "Delivering Projects."

In the meantime, DaVinci Resolve makes it easy to know if any of your project's media files are unexpectedly offline, allowing you to quickly relink them.

1 In the top left of the interface, above the media pool, click the red Relink Media button.

NOTE The Relink Media button will only be red if there are any media files in the project that are unexpectedly missing. Media that has been deliberately unlinked is not included.

2   The Relink Media window appears, telling you how many files are currently missing.

> **TIP** Place your cursor over the location where Resolve expected to find the files for more detail about the location of the files you need.

3   Click the Locate button, and in the file window that appears, navigate to the R18 Editors Guide folder and click Open.

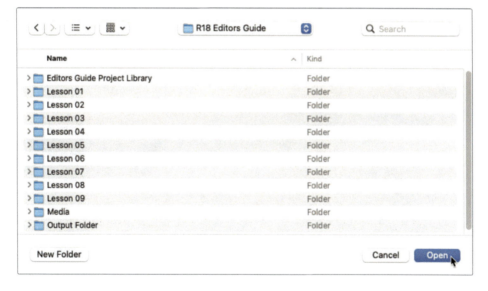

Resolve will automatically search and recognize the file structure inside this folder, and then relink the media files for you.

Once the clips have been successfully relinked, you can now begin reviewing this first project.

The Edit Page

Whether you are new to video editing, or you are already a seasoned professional looking to add DaVinci Resolve to your editing arsenal, the edit page is where you'll find all the tools you need to craft your story.

Media Pool—displays clips in the currently selected bin

Source Viewer—displays the unedited footage

Timeline Viewer—displays edited clips in the timeline

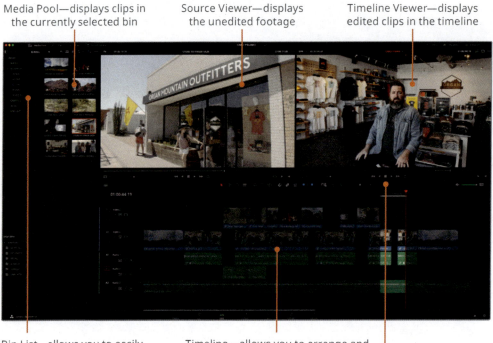

Bin List—allows you to easily select a bin to view its contents

Timeline—allows you to arrange and adjust clips used in the current edit

Timeline Toolbar—features timeline editing modes, editing functions, and timeline zoom settings

For this lesson, the project has already been set up and organized for you using a variety of different techniques that you'll explore yourself in later lessons.

Assembling the Sound Bites

Someone once said that the hardest part of writing a book is starting the first chapter. Indeed, the same is true for editing but with sounds and moving pictures rather than words on a page. Placing those first few clips into an empty timeline can be quite daunting; you never quite know where you should start, or where you will ultimately end up. However, once you have begun assembling the footage, and the edit slowly starts to reveal itself, you'll begin to see what's working, what doesn't work, and what might be coaxed into working with a bit of effort.

To start this exciting process, you'll need an empty timeline.

1   In the bin list, select the TIMELINES bin, and choose File > New Timeline or press Command-N (macOS) or Ctrl-N (Windows)

2   In the Timeline Name field in the Create New Timeline window, type **OMO Promo**, change the Audio Track Type to Mono because the audio for the interview is set to mono, and click Create.

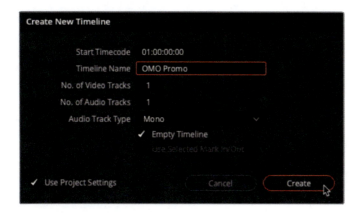

A new timeline is created in the selected bin, and additional controls appear in the timeline window.

> **NOTE** You can set the default starting timecode, number of video and audio tracks, and audio track type for new timelines in the editing section of the User Preferences.

3 In the bin list, click the disclosure triangle next to the INTERVIEW bin.

Inside this interview bin, you'll see several additional bins: one for each camera used to shoot the interview (referred to as A, B, and C cameras for the purposes of this project), one for the audio clips that have already been synced to the interview clips, and one for a series of subclips that have been created to help make working with the interview easier.

> **NOTE** You'll learn how to sync audio and video clips, create subclips, and edit multicamera footage in later lessons.

4 Select the A CAM bin to reveal the interview clips shot on this camera.

> **TIP** You can use the slider at the top of the media pool to resize the clip thumbnails, making them easier to see and making it easier to read the clip name.

To preview each of these clips, you can use the Live Media Preview feature.

5   Without clicking, move your mouse pointer over any of the four clips.

The clip appears in the source viewer and is scrubbed as you move your mouse back and forth across the clip in the media pool.

> **TIP** You can disable audio scrubbing by choosing Timeline > Audio Scrubbing or pressing Shift-S.

6   Double-click the first clip in this bin, OMO_A_002 to open it in the source viewer.

7   Click the Go To First Frame button or press Home to move the playhead back to the start of the clip.

8 Click the Play button or press the Spacebar to begin playing the clip from the beginning.

As you'll probably realize very quickly, this interview clip is rather long and encompasses several answers to different questions. You certainly don't want to use this entire clip. Instead, you will use just a small sound bite.

9 Click the Stop button or press the Spacebar again to halt playback.

To help you locate the sound bite, you can display the clip's audio waveform along with the source video.

10 Click the source viewer's Options menu (represented by three dots) and choose Show Zoomed Audio Waveform.

A green waveform appears along the bottom of the source viewer.

> **TIP** For a more detailed waveform that displays separate audio channels, you can change the source viewer mode dropdown menu to Audio Track.

11 Again, click the Go To First Frame button or press Home to return the playhead back to the beginning of the clip.

12 Click the Play button or press the Spacebar to start playing the interview again.

13 Just after Chris laughs, but before he says, "I'm Chris Lang...," click the Stop button or press the Spacebar to stop playback.

You should be able to judge where Chris starts to introduce himself from the size of the audio waveform in the display.

14 Click and hold the jog wheel (or press the Left or Right Arrow keys on your keyboard) to refine the position of the playhead (as indicated by the red line in the waveform display) to just before the start of the waveform.

> **TIP** If you disabled audio scrubbing previously, press Shift-S to enable audio scrubbing to help you refine the position to just before Chris starts speaking.

15 When you're happy that the playhead is positioned just before Chris says, "I'm Chris Lang...," choose Mark > Mark In or press I to add an In point to specify where you want this clip to start.

16 Click Play or press the Spacebar to continue playing the clip until Chris says, "...in Las Cruces, New Mexico," and then stop playback.

17 Using the jog wheel, refine the position of the playhead in the clip to just before Chris blinks, and choose Mark > Mark Out or press O to add an Out point to specify where you want this clip to end.

18 Drag the clip from the source viewer to the timeline viewer.

A series of editing overlays appears, detailing the different types of edits available to you in DaVinci Resolve. Seasoned editors will probably recognize many of these options from other nonlinear editors (NLEs), although some are specific to DaVinci Resolve. The default is Overwrite.

19 With the Overwrite edit overlay highlighted, release the mouse button.

The first clip is edited into the timeline.

Controlling Playback

An important part of editing is learning how to control the playback of your video. While you could use the transport controls underneath the source or timeline viewers, keyboard shortcuts are much more efficient. DaVinci Resolve's default keyboard layout supports all the usual shortcuts for playback that professional editors around the world will recognize. You can use the Spacebar to start and stop playback and the Left and Right Arrow keys to move forward and backward one frame at a time. More experienced users will be happy to know that the J, K, and L keys also control playback at different speeds, a staple of NLE systems.

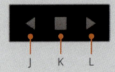

The order of the JKL keys matches the layout of the Play Reverse, Stop, and Play transport controls in both the source and timeline viewers.

Try the following to practice controlling the playback of the source or timeline viewer:

Press L to play forward.
Press J to play backward.
Press K to stop playback.

You can keep tapping the J or L keys to increase the shuttling speed up to 64x normal speed:

Press L twice to shuttle forward at 2x normal speed.
Press J twice to shuttle backward at 2x normal speed.

And you can also use the same keys to jog back and forth to precisely locate a specific frame:

Hold K and tap L to jog forward 1 frame.
Hold K and tap J to jog backward 1 frame.
Hold K and Hold L to scrub forward.
Hold K and hold J to scrub backward.

Working with the Subclips

Before you start adding additional sound bites to this timeline, it's worth considering just how much of that first clip you used. If you look at where you added the In and Out points on the clip in the source viewer, you'll see that it's only a small portion of a much larger clip—only about 6 or 7 seconds of a clip that is over 2 minutes long!

> **TIP** You can tell how long a clip is, or the duration between the In and Out points you've added, using the duration timecode value in the top left of the source viewer.

This disparity between the amount of footage shot and the amount used in the edit is not unusual and is often referred to as the shooting ratio. Depending on what's being edited, shooting ratios can vary wildly; a typical news piece might have a shooting ratio of 3:1 (for every 3 minutes shot, 1 minute was used), whereas some reality shows might have a shooting ratio of around 600:1, if not more!

With so much footage being captured and needing to be edited, it can sometimes be useful to just focus on a much smaller, relevant portion. This is where subclips come in.

Subclips are clips that have been isolated from a much longer master clip but are still referencing the original media file so that they aren't taking up additional storage space on your system. You will learn how to create and manage your own subclips in Lesson 6. For this lesson though, the subclips have already been created for you from the much longer interview clips.

1   In the bin list, select the SUBCLIPS bin to view the four subclips.

Each subclip has its own individual name that references which clip the subclip was originally created from (e.g., A 005), along with a comment on what is said in that sound bite (e.g., Design, Tagline, Brand). You will learn how to create friendly clip names like this in Lesson 6.

> **TIP** To show the filename of the clips rather than the clip name used in the project, you can choose View > Show File Names.

2   Double-click the fourth clip in the SUBCLIPS bin—SUBCLIP A 008 Experiences—to open it in the source monitor and play this clip through from the start.

Because the subclip is a much more manageable duration, it's easier to see the waveform represented over the duration of the subclip.

3   Click the source viewer's Options (three dots) menu and choose Show Full Clip Audio Waveform.

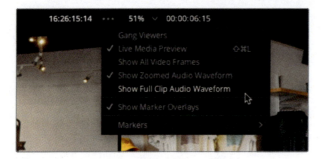

Even though this is still only a 30-second portion of the much longer interview clip in the A CAM bin, you still only need to use a portion of it.

4   Move the playhead to just before Chris starts speaking and press I to add an In point.

5   Press L to play through the clip and then stop after Chris has said "... the landscapes" but before he looks away from the interviewer, and press O to add an Out point.

6   In the timeline toolbar, click the Overwrite Clip button, or press F10.

> **NOTE** If you're using DaVinci Resolve on macOS, you may need to configure your keyboard settings in System Preferences to "Use F1, F2, etc. keys as standard function keys" to use the default editing shortcuts. Alternatively, you can use the fn key with any F key to override the macOS shortcuts.

The second interview clip is edited into the timeline starting at the position of the timeline playhead and using only the portion marked between the In and Out points in the source.

Timeline Zoom and Scroll

The edit page has three options for controlling the zoom level for clips in the timeline:

Full Extent Zoom will always display the whole duration of your timeline in the timeline window, automatically adjusting the zoom to keep everything in sight. This is most useful for seeing a bird's-eye view of your edit and allows you to navigate anywhere within the timeline.

Detail Zoom scales the timeline to a closer, zoomed view, centered on the playhead. This option is most useful when you want to step into the timeline to select a clip or edit point to make fine adjustments.

Custom Zoom provides the most flexibility since it allows you to set your own zoom scale in the timeline. You can use the slider to zoom in and out of the playhead location or hold Option (macOS) or Alt (Windows) and use the scroll function on your mouse (or trackpad) to adjust the zoom of the timeline dynamically, centered on the playhead.

Useful keyboard shortcuts for zooming the timeline include:

— Command-= (equals) in macOS or Ctrl-= (equals) in Windows to zoom in to the position of the timeline playhead.

— Command--(minus) in macOS or Ctrl--(minus) in Windows to zoom out of the position of the timeline playhead.

— Shift-Z toggles between fitting the timeline to the timeline window and returning you to the previous zoom level.

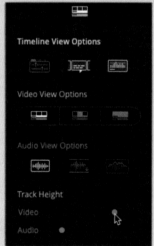

Timeline track heights can be adjusted using the Timeline View Options menu or by holding Shift and using the mouse scroll wheel over either the audio or video tracks of the timeline.

The timeline can also be scrolled left and right using the scrollbar at the bottom of the timeline window or by holding Command (macOS) or Ctrl (Windows) and using the scroll function on your mouse (or trackpad).

Adding the Final Sound Bites

You will now continue to add the rest of the sound bites for this promo using a variety of editing functions.

1   In the timeline, click the Full Extent Zoom button to have the two timeline clips fill the timeline window.

2   Select the source viewer to make it active.

> **TIP** Press Q to quickly switch between the timeline and source viewers.

3   Press the Up Arrow key to move to the previous clip in the media pool—**SUBCLIP A 008 Brand**—and, if necessary, press Home to return to the start of this clip.

4   Play the clip in the source viewer and add In and Out points around the portion of the interview where Chris says, "Our brand is really a reflection of our community who we are."

5 Perform an Overwrite edit by dragging the clip from the source viewer to the timeline viewer, clicking the Overwrite Clip button, or pressing F10.

Insert Edits

Another frequently used editing function is the Insert edit, which enables you to add clips earlier in the timeline without overwriting anything else.

1 Press Q to switch back to the source viewer and press the Up Arrow key two times to select the first clip in the SUBCLIPS bin, **SUBCLIP A 005 Design**.

2 Move the playhead in the source viewer to the second group of audio waveforms.

This is the next sound bite you will add to this timeline, but it's a bit of a tight edit to find the In point for as Chris stumbles slightly and says "that" twice. However, using the zoomed audio waveform display will make it so much easier to locate the short pause between the two "that's" quickly and accurately for a clean start to the sound bite.

3 From the source viewer's Options menu, choose Show Zoomed Audio Waveform.

4 Use the jog wheel, press the Left and/or Right Arrow keys, or hold K and press J or L, to jog the playhead into the gap between the waveforms.

5 When you're happy that the playhead is aligned after the first "that" but before the second "that," press I to add an In point.

6 Continue playing the clip in the source viewer and add an Out point after Chris says, "... that's really where the design process starts."

It doesn't make much sense for this sound bite to be edited at the end of the timeline. Instead, you will insert the clip in between two of the earlier clips in the timeline. But before making the edit, look at where the timeline playhead is currently located.

Unless you drag and drop the clip into the timeline, whenever you make an edit using the timeline viewer overlays, buttons in the timeline toolbar, or keyboard shortcuts, the timeline playhead is automatically placed at the end of the clip last edited into the timeline. This makes sense when adding clips sequentially because you can quickly edit the next clip directly after the previous clip. But if you wish to add clips to a different part of the timeline, you'll need to move the playhead to the appropriate position (see "The Rules of Three-Point Editing" later in this lesson).

Since the playhead is currently positioned at the end of the last clip in the timeline, you'll need to tell Resolve where you want to add this clip by moving the timeline playhead.

7   Press Q to switch to the timeline viewer and press the Up Arrow key to position the timeline playhead at the start of the third clip.

8   Drag the clip from the source viewer to the timeline viewer, placing it on the Insert option in the overlay.

The clip is inserted into the timeline at the playhead position.

Append At End Edits

Another useful editing function is the Append at End edit. This does exactly what it says; it adds the clip to the timeline after the last clip on the targeted track.

> **NOTE** You will learn more about targeting tracks in Lesson 2, "Refining the Rough Cut."

1   Press Q to switch back to the source viewer and press the Down Arrow key to move to the second subclip, **SUBCLIP A 007 Tagline**.

2   Add In and Out points around the final sound bite of this interview where Chris says, "That's why we say experience the southwest."

You'll want to add this sound bite to the end of the current timeline. Following the rules of three-point editing, you should position the timeline playhead at the appropriate position. However, Resolve offers a handy editing function that allows you to add clips to the end of the timeline no matter where the timeline playhead is.

3 Drag the marked subclip from the source viewer to the Append at End overlay in the timeline viewer.

The final sound bite is added to the end of the timeline even though the playhead was not placed there.

> **NOTE** If you need to catch up before moving to the next step, select the TIMELINES bin and choose File > Import > Timeline, navigate to R18 Editors Guide/Lessons/Lesson 01/Timelines and select OMO Promo Catchup 1.drt and click Open.

The "Rules" of Three-Point Editing

With a few notable exceptions, every edit you perform is generally referred to as a *three-point edit*. This means that DaVinci Resolve is calculating what you want to be edited and where you want it edited into the timeline.

In the previous steps, the In and Out points you marked in the source viewer were the first two points required; the third point was the position of the playhead in the timeline, where the In point of the clip in the source viewer will be placed when you make the edit.

Even if you don't add any In or Out points to a clip prior to editing it to the timeline, you're still following the rules of three-point editing because Resolve uses the clip in the source viewer from the beginning (the implied In point) to the end (the implied Out point).

And even if you drag a clip directly to the timeline, you are manually overriding the default third point (where the clip will start) by deciding where to place the clip prior to releasing the mouse button.

Is the Append at End edit you just performed a three-point edit? Yes! It still adhered to the rules of three-point editing by utilizing the In and Out points in the source viewer, but although the Append at End edit didn't use the timeline playhead position, it still had a third point that it was following: the end of the last clip in the timeline.

Later in this lesson, you'll learn how to make much more complex three-point edits by placing different combinations of In and Out points in the source viewer and timeline, so try to work out the rules of three-point editing that Resolve is following and how the In and Out points (real or implied) are being used to complete these edits.

In later lessons, you'll also make some four-point edits and edits that use In and Out points in unique and specific ways!

Adding B-Roll and Pacing the Sound Bites

Now that you have the sound bites in the timeline, you can start building out the story with the B-roll footage. Part of this process involves having a feel for the material you're working with and understanding how the final piece will look and feel. As with most creative processes, it requires a little imagination but also the understanding that things will continue to be refined as the edit develops.

You will start by adding a shot at the start and end of the edit, which will eventually be used as a background for the opening and closing titles.

1 Select the B-ROLL bin in the bin list.

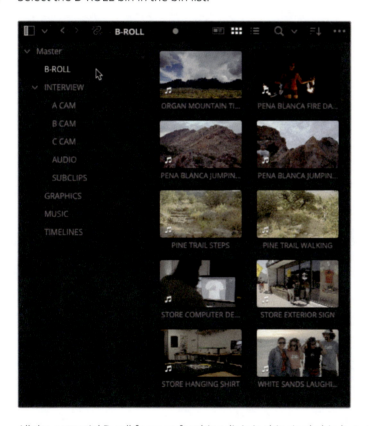

All the potential B-roll footage for this edit is in this single bin but, depending on how much footage you're dealing with, it may be difficult to see where to start. You could organize this footage further in to separate bins; however, as with the subclips you used earlier, this process has already been done for you using a series of keyword smart bins.

2   In the Smart Bins list in the bottom section of the bin list, select the Keywords folder and then select the disclosure triangle to open the list of keywords smart bins.

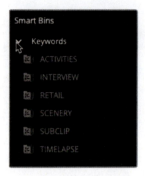

> **TIP** You can resize the Smart Bins area by dragging the dividing line at the top of the Smart Bins section of the bin list.

3   Select the SCENERY smart bin.

This smart bin contains any clips that have been identified and tagged with the "SCENERY" keyword.

> **NOTE** You will learn how to add and manage keywords and other metadata to create and manage your own smart bins in Lesson 6, "Project Organization."

4   Double-click the first clip in this smart bin, **Organ Mountain Timelapse** to open it in the source viewer.

This is a timelapse shot of the eponymous Organ Mountain and will serve well as a background to the opening and closing titles you will add toward the end of Lesson 2.

Notice that In and Out points have already been applied for you at a duration of 7 seconds.

To insert this at the beginning of your timeline, you will need to move the timeline playhead to the appropriate point.

5   Press Q to switch to the timeline viewer and press Home to move the playhead to the start of the timeline.

6   Click the Insert Clip button in the timeline toolbar or press F9.

The clip is inserted at the start of the timeline.

To add the same 7-second clip to the end of the timeline, you can perform an Append at End edit. While there is no specific timeline toolbar button for this function, you can still access the function using the menus or a keyboard shortcut.

7   Choose Edit > Append to End of Timeline or press Shift-F12.

The same 7-second clip is added to the end of the timeline.

Next, you'll create some pacing between each of the sound bites. You will start by creating a short gap after the first of Chris's sound bites.

8   Move the timeline playhead so it sits anywhere over the third clip in the timeline.

9   Choose Timeline > Select Clips Forward > Select Clips Forward on this Track, or press Y.

This command selects all the clips forward from the timeline playhead for the targeted track.

> **NOTE** You will learn more about targeting tracks in Lesson 2, "Refining the Rough Cut."

With all the clips selected, it's easy to move them together.

10 With the clips still selected, type **+200** into the timeline viewer's timecode field and press Enter (Return).

This moves the selected clips forward by 2 seconds, leaving a gap. Two seconds is an arbitrary amount that might need to be adjusted later.

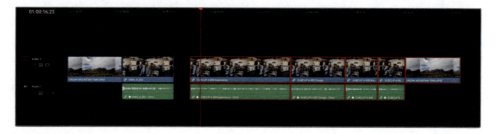

11 Press the Down Arrow key to move to the next edit in the timeline (which places the playhead on the first frame of the third interview clip) and press Y to select all clips along the targeted track.

12 Type **+100** and press Enter (Return) to add another 1-second gap between the second and third interview clips.

13 Move the playhead anywhere over the fifth clip in the timeline and press Y to select all clips forward from the playhead.

14 Again, type **+200** and Enter (Return) to have the last three clips move forward by 2 seconds leaving another gap on the timeline.

15 Finally, place your playhead over the final sound bite and press Y.

16 Type **+100** and press Enter (Return) to create a 1-second gap before the payoff of this video, where Chris recites the tagline, "Experience the southwest."

These gaps allow the sound bites to "breathe" slightly, meaning that they allow Chris to take a breath as he would have to do in the real world, but the gaps also allow room for the audience to take in what's been said before the next sound bite comes in. You can think of this technique as being the equivalent to punctuation for movie making.

Painting the Interview

Now that you have the general structure of the edit in place, you can start adding the B-roll. This performs the dual role of making Chris's sound bites come alive and covering the gaps between the sound bites, pulling the edit into a cohesive whole. This process is often referred to as painting, since you are primarily enhancing the story through pictures. To do this, you will set In and Out points in the timeline to specify the duration of the shots you'll need.

1   Move the playhead so that it snaps to the end of the second clip in the timeline, where the first gap starts.

> **TIP** Press N to enable or disable the timeline snapping function.

2   Press I to add an In point here in the timeline.

3   Play the timeline until Chris says, "... experience the southwest because..." and then stop playback.

4   Press O to add an Out point here in the timeline.

You have now marked a portion of the timeline where you want the first B-roll clip to be edited.

5   From the ACTIVITIES smart bin, double-click the clip **PINE TREE WALKING** to open it in the source viewer.

This is a shot of three friends, attired in Organ Mountain Outfitters clothing, walking in the foothills of the mountains.

6 Play the clip from the beginning and add an In point after you hear the director shout "Go ahead" and the girl is about to take her second step.

> **NOTE** Before you go to the next step, take a moment and check how many In and/or Out points you've added in total for this edit.... Three, right? The In and an Out points in the timeline are the first two, and the In point in the source viewer is the third.

Typically, B-roll shots like this tend to be edited on top of the interview already in the timeline as a cutaway. DaVinci Resolve provides an editing function to make this as easy as possible.

7 Drag the **PINE TREE WALKING** clip to the Place on Top function in the timeline viewer overlays.

Thanks to the rules of three-point editing, the clip is edited between the In and Out points in the timeline, starting at the In point you set in the viewer. The Place on Top edit has also created an extra video and audio track to accommodate the new clip without overwriting any of the existing footage.

As before, the timeline viewer is now the active window, and the playhead is automatically positioned at the end of the clip you just added to the timeline, ready for you to specify where the next edit should be.

8　Without moving the timeline playhead, press I to add an In point to the timeline.

9　Play the timeline and add an Out point after Chris says "...there's nothing like it..."

10　Press Q to switch back to the source viewer and press the Up Arrow key to open the previous shot of the friends walking up the steps, PINE TREE TRAIL STEPS.

11 Set an In point just as the second guy enters the frame and has his left leg outstretched.

12 Choose Edit > Place on Top or press F12 to add the clip to the same tracks as the previous cutaway.

13 With the timeline active, press I to add an In point in the timeline, play forward, and add an Out point after Chris says, "… ever experienced."

14 Press Q to switch to the source viewer and press the Down Arrow key until you open the shot **WHITE SANDS FRIENDS** in the source viewer.

15 Add an In point to this shot after the girl on the left looks up and is smiling for the camera.

The wind noise against the camera microphone is a little off-putting for this shot.

16 In the source viewer, click the video-only overlay and drag to the Place on Top overlay in the timeline viewer to edit only the video, not the audio of this clip.

17 Add an In point to the playhead position in the timeline and an Out point after Chris says, "… the culture, the food…"

18 Press Q to switch back to the source viewer and press the Up Arrow key to navigate to the clip **PINA BLANCA FIRE DANCER** and add an In point near the top of the clip where the girl is spinning the flaming torches.

19 In the timeline, click the destination control for A1 to disable it.

This prevents the audio from the source clip being edited into the timeline while allowing you to use editing shortcuts.

20 Press F12 to make a Place on Top edit.

21 Add an In point to the timeline and an Out point after Chris says, ".... really inspires us..."

22 Press Q to switch to the source viewer and press the Up Arrow key to navigate to the clip **PENA BLANCA JUMPING ROCKS**.

23 Add an In point when the guy is about to jump on to the rock.

24 Press F12 to make a Place on Top edit.

Hopefully, you can see just how powerful an understanding of three-point editing techniques can be to quickly add a series of cutaways like this. These cutaways will likely need trimming, but before you turn your attention to that you will add a few more cutaways to the end of the interview using a variation of the technique you've just been using.

> **NOTE** If you need to catch up before moving to the next step, select the TIMELINES bin and choose File > Import > Timeline, navigate to R18 Editors Guide/Lessons/Lesson 01/Timelines and select OMO Promo Catchup 2.drt and click Open.

Backtiming Edits

When you were adding the first set of cutaways to Chris's interview, you were specifying where each of those shots would start based on the placing of the In point. However, there are certain circumstances when you'll want to edit a clip into the timeline and specify where that shot should end. This process is often referred to as backtiming and is easy to understand when you follow the rules of three-point editing.

1   Play the third interview clip in the timeline, adding an In point just after Chris says, "...we bring it back to the store...."

To quickly add an Out point to the end of this clip, you can use a command to jump to the Out point of the clip under the playhead. This is different to simply jumping forward to the next edit because it places the playhead on the last frame of the current clip, rather than on the first frame of the next clip, so that you can add the Out point precisely.

2   Choose Playback > Go To > Last Frame or press ' (apostrophe).

The playhead jumps to the last frame of the clip, which you can see by the presence of an Out point symbol in the bottom right of the timeline viewer.

3   Press O to add an Out point.

> **NOTE** The playhead in DaVinci Resolve is inclusive of the current frame, which in practice means that In points are always added at the head, or start, of the frame, and Out points are always added at the tail, or end, of the frame. This means the minimum duration you can mark is 1 frame.

4   From the RETAIL smart bin, locate the clip named **STORE COMPUTER DESIGN** and open it in the source viewer.

This is a clip of Organ Mountain Outfitter's lead designer creating their latest T-shirt design on the computer.

5 Locate the frame, near the beginning of the clip, just before the large black circle appears.

This is where you want this shot to end.

6 Add an Out point at this frame.

Again, a bit of rudimentary math should quickly reveal that you now have a total of three In and Out points across the timeline and the source viewer (an In and Out in the timeline and another Out in the source viewer). In this case, the rules of three-point editing mean that because there is just an Out point and no In point in the source viewer, the clip will be edited to the timeline as expected, but the two Out points will be aligned, meaning the shot will then be backtimed to the In point.

> **TIP** If you need to remove an In point, you can choose Mark > Clear In or press Option-I (macOS) or Alt-I (Windows). Similarly, to remove an unwanted Out point, choose Mark > Clear Out or press Option-O (macOS) or Alt-O (Windows). To remove an In and Out point simultaneously, choose Mark > Clear In and Out or press Option-X (macOS) or Alt-X (Windows).

7 Press F12 to make a Place on Top edit.

Next, you'll add a cutaway to help bridge the gap you created in the edit earlier.

8   Add an In point to the start of the gap in the timeline.

9   Play forward and add an Out point after Chris says, "Our brand is just really a reflection of…"

10   From the RETAIL smart bin, open the clip **STORE HANGING SHIRT** in the source viewer.

This clip is a lengthy sequence of a shirt making its way out from the design studio to the shop shelves. You only need the last part of this ambitious shot though (sorry, director!)

11   In the source viewer, locate the frame where the girl hangs the T-shirt up and has left the frame (near the end of the clip).

12   Mark an Out point.

13 Press F12 to make a Place on Top edit.

14 Add an In point to the current playhead location in the timeline.

15 Add an Out point after Chris says, "That's why we say…" and he has lowered his hands to his knees.

16 From the RETAIL smart bin, open the clip **STORE EXTERIOR SIGN** in the source viewer.

17 Add an Out point on a frame just after the girl in the black hat has exited the store and releases the door.

18 Press F12 to make a Place on Top edit.

Excellent. With the final cutaway in place, all the jump cuts and gaps between Chris's sound bites have been covered. There's just one final element to add to this timeline to complete the rough cut.

> **NOTE** If you need to catch up before moving to the next step, select the TIMELINES bin and choose File > Import > Timeline, navigate to R18 Editors Guide/Lessons/Lesson 01/Timelines and select OMO Promo Catchup 3.drt and click Open.

Adding the Music

Music is such an important part of many edits. Whereas the spoken word (whether it be scripted dialogue, interviews, or narration) will often convey what we need to know about a subject, music will most often convey what we should feel about a scene or subject. Get the music wrong and the whole edit might communicate the wrong impression completely!

Thankfully, in this case the music has been carefully chosen for you. All you have to do is add it to the current timeline.

1 Ensure that the timeline viewer is selected and press Home to return the playhead to the start of the timeline.

2 Select the MUSIC bin from the bin list in the media pool and open the ONE MIN SOUNDTRACK.wav clip in the source viewer.

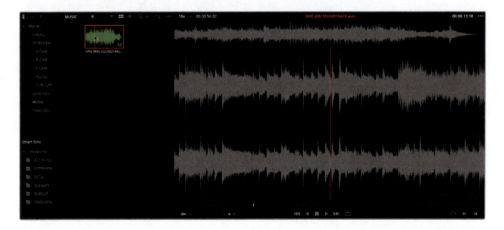

When viewing audio-only clips such as this, the source viewer automatically switches to audio mode.

You can add In and Out points to audio clips just as you've done throughout this lesson. In this case, though, it's unnecessary because the music is already just under a minute in length, which is the desired duration for the whole edit.

3   In the timeline, click the track destination control for A1 to re-enable and allow audio to be edited into the timeline again.

4   Press F12 to make a Place on Top edit.

The music clip is added to a new audio track.

The only thing left to do now is attenuate (reduce) the volume of the audio clip so that it more closely matches the rest of the audio in the timeline. It doesn't have to be the correct level at this time—just low enough that it doesn't overpower the other elements in the timeline as you continue to refine it in the next lesson.

5   Place you mouse over the volume overlay for the audio clip in the timeline, which is represented by a thin white line running through the length of the clip.

6   Click and hold the volume overlay and drag down to reduce the volume of the clip until the tooltip reads about -18 dB.

> **TIP** Hold Shift while adjusting the volume overlay for more precise control when changing the audio level.

7   Press Home to return the playhead to the start of the timeline and play back the clip to review the rough cut.

Congratulations! You have completed the first lesson. Remember, this lesson was about being able to quickly put together a rough cut using the editing tools available in the edit page. Along the way, you should have acquired a firm grasp of the principles of three-point editing. However, there is still much, much more to do to refine this timeline before it's ready to show to the client at Organ Mountain Outfitters. This will be the focus of the next lesson.

Lesson Review

1   True or false? A rough cut is a polished timeline that requires no further work.

2   Which of the following can be used to separate long clips into more manageable clips?

   a)   Source clips

   b)   Master clips

   c)   Subclips

3   True or false? The Place on Top and Append at End edit functions do not have keyboard shortcuts.

4   What is the minimum number of In and Out points you need to successfully complete a three-point edit?

   a)   0

   b)   2

   c)   3

5   What is the name given to an edit that uses a combination of two Out points and only one In point?

   a)   Reversed edit

   b)   Fit to Fill edit

   c)   Backtimed edit

Answers

1   False. A rough cut is usually the first step in the editing process where the overall structure of the edit is quickly built but still needs to be refined.

2   b) Subclips.

3   False. The default keyboard shortcut for Place on Top is F12, and for Append at End it's Shift-F12. Keyboard shortcuts for all the edit page editing functions can be seen in the View menu.

4   a) 0. Most editing functions are based on the rules of three-point editing whether or not you add In or Out points to the timeline or source viewer. If no In or Out points are in the source viewer, the first frame of the clip is treated as the In point, and the last frame is treated as the Out point. If no In or Out points are in the timeline, the playhead generally becomes the timeline In point (except when making an Append at End edit). If you drag a clip directly to the timeline, you are manually choosing the "third" point by choosing where to place the clip.

5   c) Backtimed edit.

Refining the Rough Cut

In the previous lesson, you created a rough cut for a short promotional video for the outdoor clothing brand Organ Mountain Outfitters. For many, knowing how to quickly create an edit like this is often enough. However, for many editors this is only the beginning. Now that the basic structure of the edit has revealed itself, it's now time to precisely fine-tune each individual edit so that the piece is as polished as it can possibly be.

To appropriate the 80/20 rule: the rough cut you created in Lesson 1 has accomplished about 80% of the editing required for the promo, but this should be accomplished quickly—within 20% of the available editing time. The remaining 20% of the editing (the trimming, audio mixing, graphics, etc.) will then take up the remaining 80% of the time! As you can see, the job is far from finished!

Time

This lesson takes approximately 60 minutes to complete.

Goals

Of course, part of the job of the editor is to deliver the final project to meet a deadline—you wouldn't want to miss the movie's opening night!—so it's not surprising that many feel that a job is never really "completed"; it's more that you just run out of time and money!

Setting Up the Project

This lesson starts exactly where Lesson 1 finished. If you completed Lesson 1, you may proceed to the next section in this lesson: Duplicating the Timeline.

If, however, you didn't fully complete the previous lesson, you can always import a catchup timeline to help you get started with this lesson.

> **NOTE** The following steps assume that you have at least completed the first part of Lesson 1 and that you have set up a new project library, imported the OMO Promo (Organ Mountain Outfitters promo) project file, and relinked the offline media. If you haven't completed those steps, please refer to the start of Lesson 1 before continuing with the steps below.

1   Open DaVinci Resolve and, in the Project Manager, double-click the OMO Promo project to open it in DaVinci Resolve.

2   If necessary, ensure that the edit page is selected.

3   In the bin list, select the TIMELINES bin and choose File > Import > Timeline.

4   Navigate to R18 Editors Guide/Lesson 02/Timelines, select the file OMO Promo Catchup 4.drt, and click Open.

The timeline is imported into the selected bin in your project and automatically opens in the timeline viewer. You can now continue with this lesson.

Duplicating the Timeline

It's generally good practice to duplicate your current timeline before you start making major changes because if you (or your client/director) don't like the subsequent changes you make, you always have a backup copy of the timeline to return to.

1 Choose Timeline > Find Current Timeline in Media Pool to quickly reveal the currently active timeline in its bin.

2 Right-click the current timeline and choose Duplicate Timeline.

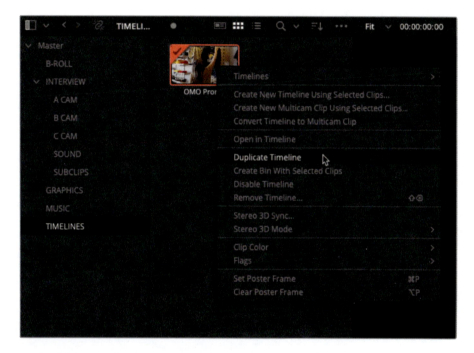

A copy of the timeline appears in the same bin.

This duplicate of your active timeline has the same name but with the word *copy* added to the end to signify this is the duplicated timeline.

> **TIP** If you cannot read the timeline's whole name in the media pool, use the slider to resize the thumbnails.

Many editors often like to rename the duplicated timeline, usually replacing "copy" with a version number (v1, v2, v3, etc.) so they know what they are looking at in the bin. However, if you leave the name of the duplicated timeline as is, subsequent duplication of the original timeline will result in the name of the new duplicated timeline being incrementally increased (copy, copy 1, copy 2, etc.). This is a useful technique since, firstly, it always means the version of the timeline you're working on is the latest, and, secondly, the automatic names of the duplicated timelines can help to "backtrack" to a previous version of the timeline if needed.

> **NOTE** Projects may contain many timelines. You can use bins to help organize these duplicated timelines so that you always know which timeline you should be working on. Alternatively, you can always disable a timeline you're not using by right-clicking the timeline in the media pool and choosing Disable Timeline. Disabled timelines cannot be opened without first re-enabling them, and they won't appear in the timeline viewer dropdown list.

With a backup copy of your timeline in your bin, you can now continue to finesse the OMO promo.

The Editor's Art

Trimming is the term given to adjusting a clip's In and Out points once it is in the timeline and is arguably the most import skill an editor possesses. Trimming allows you to adjust the start of a clip, the end of a clip, the start and end of a clip, or, in certain circumstances, the start and end of other timeline clips.

DaVinci Resolve has one of the most flexible, fully featured trimming toolsets of any nonlinear editor (NLE), allowing you to perform complex timeline adjustments intuitively and precisely. As such, over the next few lessons you will become increasingly familiar with what trimming can help you achieve in different circumstances.

Beyond simply cutting a clip and removing large sections of unwanted footage, trimming in DaVinci Resolve generally occurs in one of two timeline modes: Selection mode and Trim Edit mode.

Selection mode allows you to move clips around the timeline and adjust their durations simply and easily. This is the most intuitive way to begin trimming clips in Resolve's timeline.

Trim Edit mode unlocks the true power of the trimming functions. In this mode, you can ripple edit points, as well as slip the content of a shot and slide the position of a shot in relation to its neighboring clips.

All the trimming features in Resolve can also be applied to multiple clips or multiple edit points simultaneously and can be made by clicking and dragging with your mouse or using keyboard shortcuts for the utmost precision.

Both Selection mode and Trim Edit mode can also be used in conjunction with the Dynamic Trim mode. You'll learn more about dynamic trimming in Lesson 4, "Cutting a Dramatic Scene."

Trimming the Timeline Clips

In the previous lesson, you added a series of cutaways to paint the interview by adding In and Out points to the timeline and quickly editing the B-roll footage using the Place on Top edit. Following the steps in the lesson, you didn't spend much time reviewing each of those edits as you made them, focusing instead on just getting the material into the timeline. Now, though, you can start to consider how those shots are working together, trimming each one as appropriate.

You will begin by trimming some of the clips in the OMO Promo timeline using Selection mode.

1 Place the timeline playhead at the beginning of the first interview clip and review the first group of cutaways on the V2 track.

The edit is functional but feels a little loose, especially coming out of the interview clip into the first cutaway of the friends walking up the path in the foothills.

2 Return the playhead to the start of the first clip on V2.

3 Click the Detail Zoom button to zoom in on the playhead position in the timeline.

4 Select the clip on V2 to select it and drag it backward by about a second (-01:00 in the tooltip).

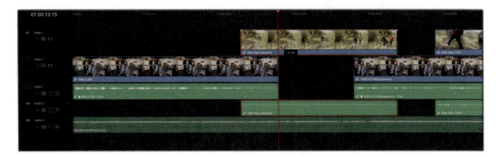

Unfortunately, moving the clip like this has left a gap in the cutaways, which disconcertingly cuts back to the underlying clip of Chris's interview on V1. You will need to trim the start of the next clip on V2 to fill this gap.

5 Click the start of the second clip on V2 and drag backward until it snaps to the end of the previous clip.

This process has lengthened the second clip by one second to fill the gap created when you moved the first clip.

6 Return the playhead to the start of the first interview clip, and play to review the changes you've just made.

Simply bringing the first cutaway in slightly earlier makes the edit feel a little "tighter," but the edit point between the first and second cutaway clips now feels a little more awkward because you've changed the point at which the second clip now starts. To refine this edit point, you will slip each shot in turn using Trim Edit mode.

7   Click the Trim Edit Mode button in the timeline toolbar, or press T.

The Trim Edit button turns red to indicate that Trim Edit is now the selected timeline mode. You will also see that the mouse pointer has changed from the arrow of Selection mode to a trim symbol.

8   Move your mouse pointer over the first cutaway clip on V2.

The Trim Edit mode is contextual, meaning that it will have different functions depending on where you place your cursor. When you place your mouse pointer over the middle of the clip, the trim symbol changes to a slip icon to reflect the type of trim you are about to perform.

9   With the slip icon displayed, click the clip and drag left in the timeline.

This time, because you are in Trim Edit mode, the clip does not move in the timeline. Instead, you will see that you are slipping the clip within its own In and Out points!

The timeline viewer has automatically changed to a 4-up multi-view preview of the change you're making.

The top two images show the start and end (the In and Out points) of the currently selected clip, and the bottom two images show the last frame of the previous timeline clip (Chris's interview on V1) and the first frame of the following timeline clip (the second cutaway on V2).

In the timeline itself, you will also see a white outline extending from the start and end of the clip being slipped.

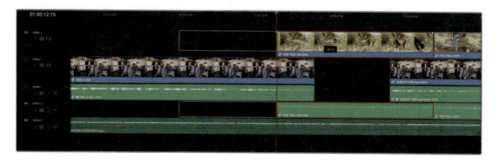

This outline shows the available handles for that clip—that is, the portion of this clip not currently being used in the timeline.

10 With the clip selected, drag to the left to slip the shot until the top-right image in the multi-view preview shows the guy in the red shirt stepping forward with his left leg forward (about -00:15 in the tooltip), and then play the first clip on V2 to review the change.

Once again, by changing this first clip, you've arguably made the cut to the second cutaway shot even worse!

11 Select the second cutaway clip and slip it to the left by about 1 second (-01:00 in the tooltip) so that the top-left image in the muti-view preview has the same guy with his left leg extended in a similar manner.

> **TIP** You can use the lower-left image (which now shows the last frame of the previous timeline clip) to help visually match the two shots.

12 Return the timeline playhead to the beginning of the first interview clip on V1 and play back to review the changes you've just made.

Even though you haven't adjusted the timing of the gap between the two interview clips on V1, by trimming and finessing the edit between the cutaway shots, the edit feels slightly tighter as a result.

> **NOTE** More often than not, you will find that you need to use different trimming operations in combination. In the previous steps, you trimmed the start of one clip but then needed to slip both clips to refine how the first shot cut to the second. As you will see, this is common to most trimming operations.

Rolling Edits

Another useful trimming function is the roll edit, which allows you to reposition an edit point by trimming two neighboring clips at the same time.

1   Position the timeline playhead at the start of the second cutaway on the V2 track, PINE TRAIL WALKING, and play the next four cutaways.

   The shot of the four friends smiling for the camera, WHITE SANDS LAUGHING FRIENDS, is a little short when viewed in context with the other cutaways.

2   Place you mouse pointer over the center of the edit, so it displays the roll icon.

3   Click to select both sides of the edit: the Out of the second clip (referred to as the outgoing clip) and the In of the third clip (referred to as the incoming clip).

4   Trim the selected edit to the right for about a second (+01:00 in the tooltip).

This rolling trim adds 1 second to the outgoing clip but also trims 1 second off the incoming clip, so it doesn't leave a gap.

> **NOTE** You can use either Selection mode or Trim Edit mode to roll edits; the functionality is the same in either mode.

5   Return the timeline playhead to the start of the WHITE SANDS FRIEND clip and review the change you've just made.

6   Now that the timing of each of the cutaways seems to work better, click the Full Extent Zoom button to zoom out and see the entire timeline.

Ripple Trimming Multiple Edit Points

Another powerful function of the timeline's Trim Edit mode is the ability to ripple edit points. Rippling edits is very useful when you want to refine the timing or pacing of shots because, unlike with Selection mode, the changes you make ripple through the rest of the timeline.

In the current edit, you can use a ripple edit to adjust the pacing of the gaps between the interview clips.

1   With Trim Edit mode still selected, click to the left of the start of the third interview clip on V1.

This will the select the "outgoing" part of the gap.

2   Drag the selected edit point to the right to begin lengthening the gap.

As you do this, you'll notice that all the other clips starting after the selected edit are also being adjusted based on the change you're making. This is the power of ripple edits.

However, the clip covering the gap between the second and third interview clips isn't included in this change because it starts before the selected edit. As a result, the clip's duration and position remain unchanged, so you will end up cutting back to Chris's interview at a different place!

While it should be no problem to simply roll the end of the cutaway of the guy on the rock, sometimes it's easier to trim multiple edit points together.

3   Release the mouse and choose Edit > Undo to undo any changes that you might have made to this edit.

> **TIP** You can view a complete list of the steps you can undo and redo by choosing Edit > History > Open History Window.

4   With the end of the gap still selected, Command-click (macOS) or Ctrl-click (Windows) the end of the fifth clip on V2.

By selecting both of these edit points, you can now trim them together.

5   Drag the selected edit on V2 to the right to add about 1 second to the duration of this clip and the selected gap below.

6   With the edits still selected, choose Playback > Play Around/To > Play Around Current Selection or press / (forward slash) to review the change.

Slide Edits

The fourth type of trim that you can make in Trim Edit mode is the slide edit. Slide edits are probably the least used type of trimming operations, but it's still useful to know that they're available to you.

Like a slip edit, slide edits are made to selected clips, but they actually affect the outgoing and incoming clips on either side of the selected clip(s).

1   Ensure that the timeline is still in Trim Edit mode and place your mouse pointer over the lower part of second of the final three middle cutaways, where you can read the name of the clip, **STORE HANGING SHIRT**.

When in Trim Edit mode, the cursor changes to the slide icon when placed over a clip's name bar.

2   With the slide icon displayed, select the **STORE HANGING SHIRT** clip and drag right, pressing N to disable snapping if necessary to slide the clip by 1 second (+01:00 in the tooltip).

> **TIP** When you disable snapping by pressing N when in the middle of adjusting a clip like this, snapping will be automatically re-enabled when you release the mouse.

3 Ripple the start of the STORE EXTERIOR SIGN clip forward by a second (-01:00 on the tooltip).

By ripple trimming the start of the clip on V2, all clips that start after this point will also be rippled. This has the effect of increasing the duration of the gap below the STORE EXTERIOR SIGN clip.

Excellent! You should now have a fuller understanding of how the Trim Edit mode functions in practice.

> **NOTE** If you need to catch up before moving to the next step, select the TIMELINES bin and choose File > Import > Timeline, navigate to R18 Editors Guide/Lessons/Lesson 02/Timelines and select OMO Promo Catchup 5.drt and click Open.

Replacing Clips

Another useful function that can help finesse the edit is using the Replace edit to quickly change an existing clip in the timeline for alternative takes or even completely different shots.

To begin with, you will use the Replace edit to change the camera angles used for the portions of Chris's interview that are still in view. You will start with the last interview clip, which you will replace with a close-up shot.

> **NOTE** The following examples use clips that have matching timecode to aid the replacing of clips. This is not the same as using the full multicam editing workflow. You'll explore how to sync, cut, and switch between the different angles of multicamera footage in Lesson 5, "Multicamera Editing."

1 Place the timeline playhead so that it snaps to the end of the cutaway over the final interview clip.

2 In the timeline viewer, right-click the current timecode field and choose Source Timecode.

The current timecode field now displays the source timecode for the currently displayed clip in the timeline.

3 Right-click the source timecode field and choose Copy Timecode.

4 In the media pool, select the B CAM bin.

5 From the name of the interview clip in the timeline, SUBCLIP A 007 Tagline, you can identify that this sound bite comes from take "007" of the full interview.

6 Open the clip OMO_B_007 into the source viewer.

7 Right-click the source viewer's current timecode field and choose Paste Timecode.

The playhead jumps directly to the pasted timecode in the source viewer.

TIP You can see the current source timecode for each clip at the current playhead position by selecting the timeline viewer's Options menu (three dots menu) and choosing Show Timecode Overlays.

While these overlays are for reference only and cannot be copied directly, you can use them to check the source timecode for the clip on a specific track and then manually type the timecode in the source viewer's current timecode field.

Before you go further, note that the Replace edit is a powerful editing function, but it does not follow the standard rules of three-point editing. The Replace edit uses the position of the two playheads in the source and timeline viewers to determine how the source clip should be aligned in the timeline and, in the absence of any timeline In or Out points, will use the timeline clip's start and end points to determine the duration of the clip.

It can be a little confusing at first, but remember that the Replace edit is the first of the *four-point edits* you can perform in DaVinci Resolve's edit page: the two playheads being the first two points and the timeline clip's start and end points being the second two points.

8   Drag the clip from the source viewer to Replace in the timeline viewer overlays.

The close-up from the B camera replaces the wide shot in the timeline, with the source and timeline viewers now both displaying the same frame.

9   Press / (forward slash) to preview the new clip in the timeline.

As if by magic, the audio and video of the sound bite clip you originally edited into the timeline has been replaced with the audio and video of the slightly closer camera angle! Such is the power of the Replace edit.

Replacing Part of a Clip

You can also use the Replace edit to replace a portion of a clip in the timeline in a couple of ways.

1. Place the timeline playhead in the middle of the third interview clip, just after Chris says, "...we take that inspiration...."

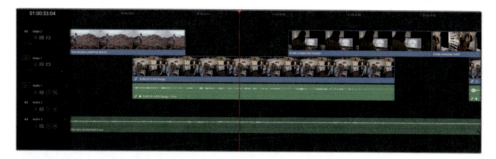

2. In the timeline toolbar, click the Blade Edit mode or press B.

3. Place the mouse pointer over the interview clip so that the Blade mode snaps to the timeline playhead.

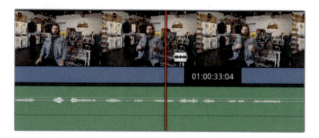

4. Click the clip to add an edit point.

5 Click the timeline viewer's timecode field and press Command-C (macOS) or Ctrl-C (Windows) to copy the source timecode for the interview clip.

6 From the C CAM bin, open the **OMO_C_005** clip in the source viewer.

7 Select the timecode field in the source viewer and press Command-V (macOS) or Ctrl-V (Windows).

8 Click the Replace Clip button in the timeline toolbar or press F11 to replace the timeline clip after the edit.

> **NOTE** On macOS systems, you might need to make a further change to the default keyboard functions in order to use F11 as the shortcut for Replace edits. Open System Preferences and choose Keyboard > Shortcuts > Mission Control, and either deselect or change the shortcut used for Show Desktop.

9 Press / (forward slash) to preview the edit. If necessary, select Trim Edit mode and roll the edit point to refine the position of the cut.

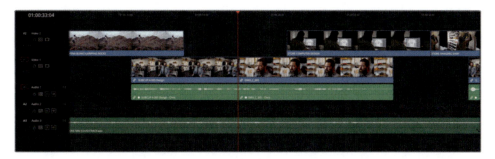

You can also use timeline In and Out points in conjunction with the Replace edit to limit the amount of footage replaced in the timeline, and use the track destination controls to specify whether to replace the video or audio parts of a clip.

10 Press A to select Selection mode or click the Selection mode button in the timeline toolbar.

11 Play the first interview clip and add an In point just before Chris says, "… store in Las Cruces…."

12 Choose Playback > Go To > Last Frame, or press ' (apostrophe), to jump to the last frame of this clip and add an Out point.

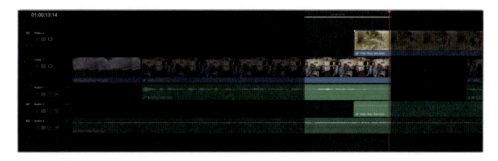

Unfortunately, you can't simply copy the source timecode as you did previously because the cutaway clip is being displayed—so the source timecode is actually the source timecode for the PINE TRAIL WALKING clip. Instead, you will perform a Match Frame to open the interview clip's source clip in the source viewer.

13 Select the first interview clip in the timeline and choose Clip > Match Frame to Source Clip or press F.

> **NOTE** Alternatively, you can select the Match Frame button to the right of the Timeline viewer's controls.

The source clip for the interview clip in the timeline opens in the source viewer at the same frame as the playhead in the timeline.

14 Select the current timecode field in the source viewer and press Command-C (macOS) or Ctrl-C (Windows) to copy the current timecode.

15 From the C CAM bin, open the OMO_C_002 clip into the source viewer and paste the copied timecode into the current timecode field.

Because you only want to use the video of this clip, you need to specify that the audio shouldn't be used to replace the audio already in the timeline.

16 In the timeline track headers, click the destination control for source audio A1, or press Command-Option-1 (macOS) or Ctrl-Alt-1 (Windows).

17 Press F11 to make the Replace edit, replacing only the video part of the clip in the timeline.

Placing In and Out points in the timeline allows you to limit the amount of footage being replaced. By using the additional camera angles of the interview, you have also introduced more visual variety to the edit.

Specifying Destination Tracks

In the previous lesson, you started by editing a series of clips on to the first video and audio tracks in the timeline: Video 1 and Audio 1. The clips you subsequently added as cutaways were edited on to additional tracks (Video 2 and Audio 2, and then Audio 3 for the music) using the Place on Top edit, which automatically created those additional tracks as they were needed.

However, if you wish to edit directly to an existing track, you will need to specify that track using the destination controls in the timeline track headers.

1   Place the timeline playhead at the start of the fifth clip on V2, **PENA BLANCA JUMPING ROCKS**, and add an In point.

2   Add an Out point about 2 seconds later, as the guy reaches the top of the rock.

3   From the ACTIVITIES smart bin, open the clip **PENA BLANCA LOOKOUT** into the source viewer and locate a frame about halfway through the clip that most closely matches the frame in the timeline viewer.

You now need to target the appropriate track in the timeline.

4 In the timeline track header, click the V2 destination control to target the video source (V1) to the Video 2 track, or press Option-2 (macOS) or Alt-2 (Windows).

> **TIP** A series of commands for changing the various video and audio destination controls can be found by choosing Timeline > Track Destination Selection.

5 Press F11 to perform a Replace edit directly on the Video 2 track.

6 Press / (forward slash) to preview the edit you've just made.

An advantage of using the Replace edit, rather than a backtimed Overwrite edit, is that you don't need to add any In or Out points in the source viewer. Indeed, any In or Out points in the source viewer will be ignored whenever you make a Replace edit. Also, when you perform Replace edits, the source and timeline playheads can be placed on frames outside the In and Out points in the timeline, but the portion of the source footage replaced within the In and Out points will be calculated from the offset of the In and Out points from the playheads.

7 Right-click the timeline viewer's timecode field and choose Record Timecode to return to displaying the timeline's timecode value.

NOTE If you need to catch up before moving to the next step, select the TIMELINES bin and choose File > Import > Timeline, navigate to R18 Editors Guide/Lessons/Lesson 02/Timelines and select OMO Promo Catchup 6.drt and click Open.

Audio Mixing

So far in this lesson, you have primarily continued to refine the video clips in the timeline, and the nature of editing means that you can continue making similar adjustments as you see fit, whether it be replacing shots entirely or in part, or trimming clips to refine the pacing, timing, and duration of the edit. However, at some point you need to switch your focus to other parts of the edit that need your attention, especially if you're going to deliver the edit to the client in good time!

As the edit seems to be progressing well, it's time to turn your attention to the audio in the timeline.

In many cases, you can spend as much time refining and finessing the audio in the timeline as you can the video. Over the next steps, you'll use some common techniques to ensure that you can mix your audio to the correct levels efficiently and effectively. You will start by normalizing Chris's interview levels.

NOTE You'll learn more advanced audio editing and mixing techniques in Lesson 8, "Audio Editing and Mixing."

1   Click the Mixer button at the top right of the interface to display the mixer to the right of the timeline.

TIP If you're struggling for room on your display, click the Media Pool button in the top left of the interface to close the media pool and reclaim some screen space.

2   Click the Mute button for track A3 in the timeline track controls or the Mixer so the music doesn't play. This lets you to focus on the audio levels of Chris's interview clips.

3   Play the timeline from the start, observing the levels of the clips on track A1 in the mixer.

Even though the audio for these clips is from the same interview, there is still a lot of inconsistency in the levels, with some clips peaking as loud as -3 dBFS and others peaking as low as -18 dBFS.

4   Select all the clips on the A1 track, and then right-click any of the selected clips and choose Normalize Audio Levels.

The Normalize Audio Level window appears with several options for normalizing the audio levels of the selected clips.

5   Leave the Normalization Mode set to Sample Peak Level and the Target Level set to the default of -9 dBFS.

The Normalization Mode dropdown menu allows you to choose the method used to determine how each clip's volume level will be normalized. Options include a variety of loudness normalization algorithms specific to various international standards, which are useful for balancing the perceived overall loudness of several clips to one another, regardless of transient levels throughout each clip. You can also perform Peak normalization, with options for both Sample Peak and True Peak. Whichever method you choose, the result is largely the same: each clip's audio level is analyzed and then adjusted to the various peak and/or loudness levels specified.

6   Change the Set Level option to Independent.

When Set Level is set to Relative, all selected clips are treated as if they're one clip, so that the highest peak and/or loudness level of all the selected clips is used to define the adjustment, and the volume of all selected clips is adjusted by the same amount. When Set Level is set to Independent, the peak and/or loudness levels of each clip is used to define the adjustment to that specific clip. This is likely to result in different volume adjustments to each clip to make the peak and/or loudness levels of each audio clip better match one another. Relative is useful if you're normalizing a series of clips that have a consistent recorded level, such as a controlled dialogue recording, whereas Independent is much more useful if you're trying to balance a series of clips that have different recorded levels (subtle or not), such as interviews or other location audio, which might have been recorded under less-controlled conditions.

7   Click Normalize.

The audio levels for the selected clips are adjusted so that the peak level for each clip hits -9 dBFS on the audio meters.

8   To verify, play through the timeline again, looking at the levels of the A1 track in the mixer.

Each clip now plays at a more consistent level but not necessarily completely consistently because it depends on the consistency at which the original audio levels were recorded! For example, the first and fifth interview clips are both generally around 6 dB lower than the other clips. This is because both clips have a larger peak at their starts, which affects how the rest of the clip has been normalized.

9   Zoom in on the first clip on A1 so you clearly see the audio waveform.

10 Option-click (macOS) or Alt-click (Windows) the volume bar twice to add two keyframes just after the first large peak, but before the second lower peak.

11 Click and hold the volume bar after the second keyframe to see the current level adjustment applied to the clip by the Normalization process.

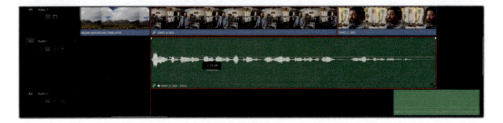

The tooltip shows an adjustment of about -3 dB. To bring the level of the rest of this clip back in line with the other clips, you need to apply a +6 dB adjustment, bringing the adjustment up from -3 dB to +3 dB.

12 Drag the volume bar up until the tooltip reads about 3 dB.

13 Repeat the process on the fifth clip on A1, SUBCIP A 008 Brand - Chris, adding the keyframes just after the first large peak, and bringing the level after the second keyframe up to around 4 dB.

Congratulations! You have successfully normalized and balanced the interview clips.

Using Keyframes to Remove Unwanted Audio

You can also use audio keyframes to help you remove unwanted sounds.

1 Play the last interview clip on track A1, OMO_B_007, listening carefully after Chris says, "That's why we say...."

You can hear an unwanted clapping sound as he lowers his hands to his knees (this action is covered by the cutaway of the store exterior).

2 Zoom in on the clip in the timeline so you can clearly see the waveform and identify the slight peak created by the clap.

3 Option-click (macOS) or Alt-click (Windows) the volume bar to add two keyframes just after Chris's first sound bite and just before the second, and a third above the peak made by the clap.

4 Drag the middle keyframe down to reduce the level of the clap so that you no longer hear it in the mix.

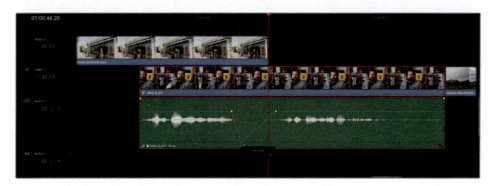

Setting Levels of SOT Clips

You can also apply normalization to effects clips, such as those on track A2. This type of audio is often referred to as SOT, short for "sound on tape," an anachronistic term used to refer to audio recorded on a camera.

1   Select the two clips on track A2, right-click either of them, and choose Normalize Audio Levels.

2   In the Normalize Audio Level window, change the Target Level to -18 dBFS, a level more in keeping with sound effects, or SOT clips, and click Normalize.

The clips are normalized to the desired level.

3   Add two keyframes to the first of the SOT clips, just before Chris's second interview clip, and raise the first part of this clip by about 6 dB (from about 12 dB to 18 dB in the tooltip), so you can hear more of the walkers' footsteps between Chris's interview clips.

4   Use the fade-in handle to fade in the first SOT clip over about a second.

5 Use the fade-out handle to fade out the second SOT clip over about a second.

These last few changes mean the SOT clips are now working well with the audio from Chris's interview. Next, you'll need to bring the music back into the mix.

"Ducking" the Music Levels

The final touch you will apply to the audio for this edit is to mix in the music with the rest of the timeline audio, reducing the volume level of the music during interviews or other dialogue clips—a technique referred to as ducking.

The music's audio level is currently set at -18 dB, which is a good starting point for the music in relation to the levels of the interview clips. Having already set these levels, you'll want to simply raise the audio in the gaps between Chris's sound bites.

1 Click the Mute button for A3 in the timeline track controls or the Mixer to re-enable the audio for the music, and play the timeline from the beginning, listening to how the music sounds next to the interview audio.

2 Add two keyframes to the music clip near to where Chris's first sound bite begins.

3 Drag the volume bar before the first of these two keyframes up by about 6 dB (to about -12 dB in the tooltip).

Play the beginning of the timeline again to hear the music duck under Chris's interview audio.

4   Repeat the process, adding two keyframes to the music clip at the end of Chris's final sound bite, increasing the level of the clip after the second keyframe by about 6 dB so that the music swells after Chris recites the tagline for this video: "...experience the southwest." Play back this part of the timeline to review the change.

There are several other gaps in Chris's interview where you can also increase the level of the music.

5   Hold Option (macOS) or Alt (Windows) and click to add four keyframes to the music clip around the gap between Chris's first and second interview clips.

6   Drag the volume bar between the second and third of these keyframes to raise the level of that part of the audio clip by about 6 dB.

7   Repeat the process to raise the music during the other gaps between the interview clips.

You can now review the timeline from the beginning to hear the music duck under Chris's sound bite clips. Once you're happy with each of the clips' levels, you can also close the Mixer window.

Adding the Logo

The edit is nearly finished. Just before you apply the finishing touches though, you'll need to add the Organ Mountain Outfitters logo to the opening shot and some titles to the closing shot featuring a "call to action" to encourage the viewer to visit the Organ Mountain Outfitter's website as well as emphasizing the tagline, "Experience the Southwest."

1   In the timeline, position the playhead over the first clip, **ORGAN MOUNTAIN TL.mov**, and then choose Playback > Go To > Last Frame or press ' (apostrophe) to jump to the last frame of this clip.

2   Press O to set an Out point at this frame in the timeline.

3   From the GRAPHICS bin, open the clip **OMO LOGO.png** in the source viewer.

4 Click the Overwrite Clip button in the timeline toolbar, or press F10, to overwrite the clip on to the targeted track (V2).

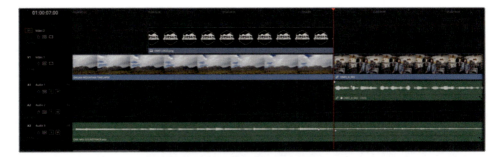

DaVinci Resolve can work with various graphic file formats as well as video and audio files. When using graphic files, the clips have a default "duration" of 5 seconds, which was enough for the backtimed edit you just performed in the previous step. However, because graphic files are simply the same frame repeated, the clip can be trimmed to be as long or as short as you need it.

5 In the timeline, drag the fade handle at the start of the clip to the right to apply a 12-frame fade in (+00:12 in the tooltip).

6 Place the timeline playhead over the **OMO LOGO.png** clip and in the top left of the interface, click the Effects button to open the Effects Library below the media pool.

The Effects Library contains several built-in video and audio transitions, titles, and generators.

7 Select the Open FX category and scroll down to the Resolve FX Stylize group.

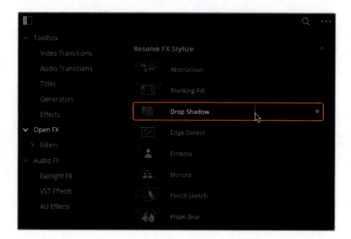

8 Double-click the Drop Shadow filter to apply the effect to the **OMO LOGO.png** clip in the timeline.

The drop shadow helps the logo stand out from the background clip on V1. To adjust the settings for this clip, you will need to use the controls available in the Inspector.

9   Click the Inspector button in the top right of the interface to open the Inspector to the right of the timeline viewer.

The Inspector is the place where you can access many settings for the clips in the timeline or the media pool. You will learn how to use the Inspector in later lessons. For now, all you need to do is adjust the Drop Shadow effect's settings.

10 In the Inspector, click the Effects tab to reveal the controls for any effects applied to the currently selected clip.

11 Reduce the Drop Distance to about 0.020 and the Blur to about 0.40 to create a more defined shadow.

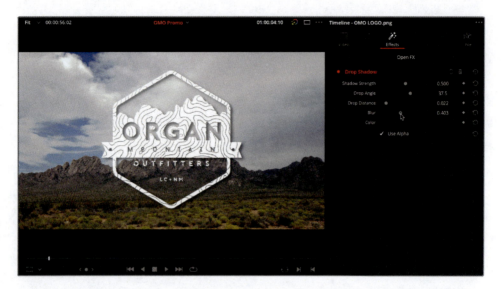

With the drop shadow applied and finessed, there's one other useful effect you can apply to this graphic to provide some much needed motion.

12 In the Inspector, click the Video tab.

This tab in the Inspector allows you to adjust many common settings for the current clip, including size and position. You will learn how to use these controls to create eye-catching effects in Lesson 7, "Compositing in the Edit Page."

13 Enable the Dynamic Zoom and click to reveal the controls and change the Dynamic Zoom Ease menu to Ease Out.

Play the **OMO LOGO.png** clip in the timeline to review the dynamic zoom results, and notice how the still image fades and zooms, coming to a gentle rest toward the end of the clip.

You can also choose to refine the start and end framing for the dynamic zoom.

14 Choose View > Viewer Overlay > Dynamic Zoom to reveal the onscreen controls for the dynamic zoom.

> **TIP** You can also reveal the controls in the viewer overlay by clicking the Transform Mode dropdown menu in the timeline viewer.

15 Adjust the starting framing for the dynamic zoom by dragging the corners of the green box out slightly, away from the edges of the graphic.

16 Once you're happy with the starting and ending framing of the dynamic zoom, choose View > Viewer Overlay > Toggle On/Off or press Shift-` (grave accent) to turn off the viewer overlays.

Excellent. By utilizing some of the built-in effects and controls in the Inspector, you have taken a simple still image and used it to create an eye-catching opening for the promo. You'll learn more about how you can composite images in the edit page in Lesson 7.

Adding the Closing Titles

Next, it's time to add the call to action using one of Resolve's built-in Fusion Titles templates.

1 Scroll to the end of the timeline and play the final clip, **ORGAN MOUNTAIN TIMELAPSE**. Using the audio waveforms as a guide, stop when you hear the final beat of the music.

2 In the Effects Library, select the Titles category and scroll through the list of Fusion Titles to the **Horizontal Line Reveal** title.

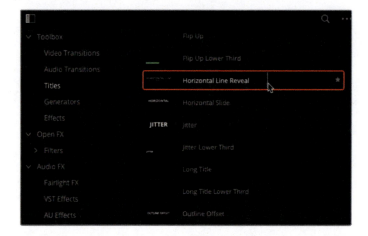

> **TIP** You can live preview each of the Fusion Titles.

You can't open a title in the source directly from the Effects Library; you must edit them directly to the timeline. But that doesn't mean you can't use three-point editing techniques!

3 Drag the **Horizontal Line Reveal** title from the Effects Library to the timeline viewer to perform an Overwrite edit to the targeted video track (V2).

> **NOTE** If your timeline is currently targeting the V1 track, you should use the Place on Top edit.

4 Place the timeline playhead over the title clip in the timeline.

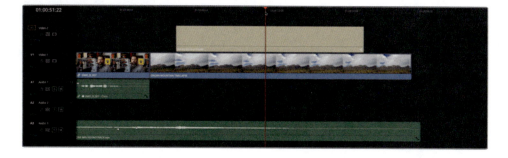

5   In the Inspector, select the SAMPLE UPPER text in the Upper Text Controls and type **organmountainoutfitters.com**.

6   Choose View > Safe Area > On.

The Safe Area is used to ensure that titles and graphics are properly displayed on screens that have an overscan and is typically a requirement for broadcast programs.

7   Change the Upper Text Size to about 0.08 so the title fits within the inner title safe area.

8   Scroll down in the Inspector to the Lower Text Controls, highlight the SAMPLE text and type **#experiencethesouthwest**.

9   Adjust the Lower Text Spacing to about 1.09 so the text is about the same length as the line.

To make the text more eye-catching, you can adjust the colors.

10   Scroll back up to the Upper Text Controls.

11   Change the Type 1 option to Gradient.

12 In the Shading Gradient 1, click the left triangle control and change the color to a rich orange.

13 Click the right gradient triangle and change the color to warm yellow.

14 Scroll to the bottom of the Inspector to the Line Color controls.

15 Change the Type menu to Gradient to reveal similar controls.

16 Click the left gradient control and adjust the color to a deep orange.

17 Select the right gradient control and adjust the color to a warm yellow.

18 Click the Settings tab at the top of the Inspector.

19 Change the Y Position value to about -360.00 to move the title down in the timeline viewer but still inside the inner title safe area.

20 In the timeline, use the title clip's fade handle to apply a 12-frame fade out (-00:12 in the tooltip).

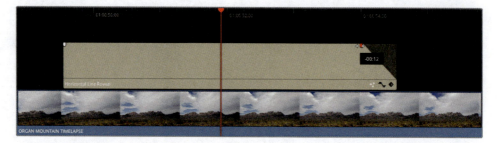

21 Trim the end of the ORGAN MOUNTAIN TL.mov clip on V1 so it snaps to the end of the audio clip on A3 and apply a 1-second fade out (-01:00 on the tooltip).

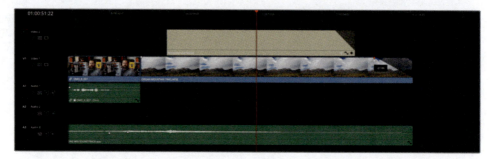

22 Choose View > Safe Area > On to toggle the safe area overlay off.

> **NOTE** If you need to catch up before moving to the next step, select the TIMELINES bin and choose File > Import > Timeline, navigate to R18 Editors Guide/Lessons/Lesson 02/Timelines and select OMO Promo Catchup 8.drt and click Open.

Last Minute Changes

The edit for the OMO Promo is very nearly complete and nearly ready for the client to review. In fact, you've just got the call that they are on their way to you now! But with a few minutes to spare, you can make some final adjustments.

Stabilizing Shots

You will start by stabilizing the shot of the guy on the rocks.

1   In the timeline, play the second clip of the guy standing on the rock looking out over the mountains, **PENA BLANCA JUMPING ROCKS**.

This clip is looking great, but the shake from the handheld camera is quite noticeable and is detracting from an otherwise great shot.

2   Place your playhead over the clip in the timeline and, in the Inspector, open the Stabilize controls and click the Stabilize button.

Resolve analyzes the clip and attempts to stabilize the shot. Once the analysis has completed, play the shot to review the changes.

The shot seems to be a little less shaky, but it doesn't completely smooth the camera movement.

3   Increase the Smooth control to about 0.900 and click the Stabilize button again to apply the changes.

The increased smoothing value helps to reduce the camera shake even further, resulting in a much-improved shot.

Changing Clip Speed

Another quick adjustment you can apply is to change the speed at which a shot plays back.

1   Place the timeline playhead over the clip **PENA BLANCA FIRE DANCER** and, in the Inspector, open the Speed Change controls.

2   Change the Speed % value to 50.00 to play back the clip at 50%.

> **NOTE** You'll learn how to make more sophisticated speed changes in Lesson 7.

Adding Transitions

Finally, just as you hear the client's footsteps approaching, there's just enough time to add some video transitions to the final clips.

1 In the timeline, select the edit point between the last interview clip and the **ORGAN MOUNTAIN TIMELAPSE** clip.

2 In the Effects Library, select the Video Transitions group and double-click the Cross Dissolve transition to apply it to the selected edit point.

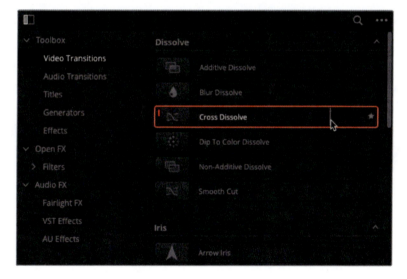

3 Select the transition in the timeline and press Command-D (macOS) or Ctrl-D (Windows) to open the Change Transition Duration window.

4   Click the Change Duration field and type **12**, and then click Change to change the
    duration of the selected transition to 12 frames.

5   Press Command-C (macOS) or Ctrl-C (Windows) to copy the selected transition.

6   Press T to switch to Trim Edit mode and click and drag a selection window across the
    two edit points between the final three cutaway clips on V2 to select them.

7   Press Command-V (macOS) or Ctrl-V (Windows) to paste the copied transition to the
    selected edit points.

    Just in time! The client has arrived with lattes in hand and is taking their place, ready
    for the big screening!

Full Screen Review

Congratulations! You have successfully edited a short but polished promotional clip for
Organ Mountain Outfitters. However, before you deliver the final file that will be uploaded
to their social media channels, there's an opportunity to watch the whole thing back at full
screen. This can be very useful because it will allow you (and the client) to see the results of
your efforts without being distracted by the "editing paraphernalia" of the interface.

1   Choose Workspace > Viewer Mode > Cinema Viewer or press Command-F (macOS) or Ctrl-F (Windows).

DaVinci Resolve displays the current timeline in full screen. Simple onscreen navigation and playback controls are available as an overlay.

> **TIP** You can still use the keyboard shortcuts you're familiar with to navigate around the timeline in the Cinema Viewer. Use Home to return to the start of the edit, J, K, and L for playback, etc.

2   Use the onscreen controls to return to the start and begin playback. The controls and your mouse pointer will disappear after a few seconds.

3   When you've finished watching your masterpiece, move the mouse slightly to display the controls again and click the Exit Fullscreen button, or press Esc (Escape), to return to the full Resolve interface.

Playing back your timeline like this gives you an opportunity to see your edit the same way as your viewers will. Watch carefully and see if there are any parts of the edit that might benefit from additional changes. If so, now is the time to make those adjustments.

> **NOTE** If you need to catch up before moving to the next step, select the TIMELINES bin and choose File > Import > Timeline, navigate to R18 Editors Guide/Lessons/Lesson 02/Timelines and select OMO Promo Catchup FINISHED.drt and click Open.

Quick Export

The client loves the edit and is happy to sign off on your work. Now, the only step left is to export the edited timeline so it can be posted to popular streaming and social media websites as quickly as possible. You can accomplish all this quickly using DaVinci Resolve's Quick Export feature.

1 Choose File > Quick Export.

The Quick Export window includes commonly used presets for creating a video file of your currently open timeline, including uploading directly to common social media and sharing sites including YouTube, Vimeo, Twitter, or Dropbox.

> **NOTE** DaVinci Resolve Studio provides additional options for sharing via the video review services Dropbox Replay and Frame.io.

2 Select the YouTube option.

This preset contains all the settings needed to create a file and upload it directly to YouTube. However, if you haven't already input your account information, you'll just see a Manage Account button.

3 Click the Manage Account button if you haven't already input your account information for YouTube.

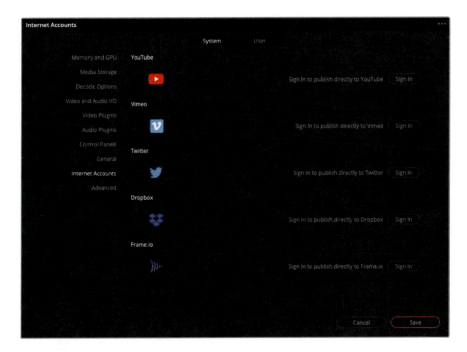

Clicking the Manage Account button opens the DaVinci Resolve System Preferences with the Internet Accounts panel selected where you can enter account information to allow DaVinci Resolve to upload directly to these web services on your behalf.

> **TIP** You can also access these preferences by choosing DaVinci Resolve > Preferences > Internet Accounts.

To sign in to any of your accounts for these sharing sites, click the Sign In button for the appropriate service and follow the directions to enter your account information and give DaVinci Resolve the relevant permissions. When you've completed the sign-in process, the Quick Export window will now show an Upload Directly checkbox. Selecting this allows you to enter a title, privacy settings, and a description.

Instead of uploading directly to a video sharing service, though, you might just want to create a stand-alone file. You can then manually upload the file instead or distribute it in other ways.

4   Click Cancel to return to the Quick Export window and select the H.264 Master preset.

5   Click Export, choose a location where you want the movie to be saved on your computer, and then click Save.

> **TIP** By default, the exported file will have the same name as the timeline you're outputting, but you can also rename the exported video at this stage if necessary.

A render progress window shows the time to completion and, if appropriate, the upload progress for the movie. Once the render progress bar has completed, close the Quick Export progress window and you will find the exported video in the location you chose for it, so you can open it in your computer's default video player.

Congratulations! Over these first two lessons, you have successfully put together a short, yet complex, promo using the editing toolset available to you in DaVinci Resolve's edit page. Hopefully, these lessons have given you some insights into how these tools function and how you can start to use them in your own work. However, this is just the beginning of your editing journey. Throughout the rest of this book, additional lessons will help you explore how to use DaVinci Resolve to edit a variety of genres—from interviews and dramatic footage to complex multicamera events.

For the moment though, take a breath….and enjoy that latte.

Lesson Review

1   Which timeline mode(s) allow you to trim the start or end of a clip in the timeline?

   a)   Selection mode

   b)   Trim Edit mode

   c)   Blade Edit mode

2   Which timeline mode(s) allow you to slip a clip in the timeline?

   a)   Selection mode

   b)   Trim Edit mode

   c)   Blade Edit mode

3   True or false? The Replace edit uses the position of the timeline and source viewer playheads but always ignores In and Out points in the timeline.

4   Which modifier key is used to add keyframes to a clip's volume bar in the timeline?

5   Where do you sign in to supported video sharing services to be able to upload your video automatically?

   a)   Quick Export window

   b)   Sharing tab in the Inspector

   c)   Internet Accounts in System Preferences

Answers

1 a) and b). Selection and Trim Edit modes can be used to trim the start and end of a clip in the timeline.

2 b). Trim Edit mode allows you to slip a clip in the timeline by adjusting the In and Out point of the clip at the same time.

3 False. The Replace edit will use In and Out points in the timeline to limit the amount replaced, but will always ignore any In or Out points in the source viewer.

4 Option-click (macOS) or Alt-click (Windows) will allow you to add a keyframe to an audio clip's volume bar in the timeline.

5 c). You sign in to supported video sharing services using the Internet Accounts section of the System Preferences, although you can access this part of the UI from the Quick Export window.

Fine Cutting an Interview

Getting the very best out of your interviewee takes a lot of practice and a good ear for the spoken word.

When cutting interviews, it's common to initially create the best-sounding interview and almost completely disregard the visuals. This cut is commonly called a radio edit because it's similar to editing an audio-only interview. Once you have the interview audio cut properly, you can turn your attention to the video edits, often referred to as the paint because you are illustrating, or "painting," your interview with appropriate pictures from your B-roll rushes.

In this lesson, you'll use advanced audio and video trimming and other DaVinci Resolve workflow features to finish an engaging social media promo piece for a 100% vegan restaurant, Miss Rachel's Pantry.

Time

This lesson takes approximately 50 minutes to complete.

Goals

Starting the Project

You'll begin this lesson by importing and opening a DaVinci Resolve project that has most of the radio edit already edited but includes some clips in which the audio hasn't been examined for clarity. You'll play the clips and then begin to identify and remove the small stutters and stray "umms" to get the best sound bites from the interviewee.

1   In the Project Manager, right-click and choose Import Project.

2   Navigate to R17 Editing Lessons > Lesson 03 and select **WELCOME TO THE PANTRY.drp**.

3   Double-click the imported project in the Project Manager.

4   If necessary, click the Edit page button.

5   Click Relink Media in the Media Pool and relink the offline clips to their media files in the R18 Editors Guide folder.

6   Choose Workspace > Reset UI Layout to reset the workspace.

7   Resize the timeline tracks and click the Full Extent Zoom button to see the entire timeline.

   This timeline is the first part of the promo and has already been cut for you.

8 Return your playhead to the start of this timeline and play through to review the edit so far.

This piece is being driven by an interview with the eponymous Miss Rachel, owner and head chef at Miss Rachel's Pantry. Notice how the interview has been edited so that it sounds clean, without many distracting pauses, stutters, or poorly chosen words.

> **NOTE** Feel free to review the rushes from the Interview Clips bin in the Rushes bin to get a sense of how this edit was pieced together from the original clips.

You will also notice the gaps that were left intentionally to add short pauses between thoughts or subjects and to allow the edit to "breathe." Small audio clips of wildtrack or atmos help to fill the gaps left in the audio with the ambient sounds of the restaurant. Much of the distracting jump cuts have been covered by editing B-roll footage onto the V2 track.

Two red markers are also present in this timeline referencing certain locations.

> **TIP** You can add your own markers to any timeline by clicking the Markers button in the timeline toolbar or by pressing M. Double-click an existing marker to edit its color, name, or other properties.

To get a sense of how the cut interview would look without the cutaways, you can disable the V2 track.

9   In the timeline track header for the Cutaways track (V2), click the Disable Video Track button.

All clips on the V2 track are disabled.

10   Return the playhead to the start of the timeline and play through the edit once again.

Wow! The edited interview alone isn't quite so polished. The jump cuts between the edited sound bites really distract from what Miss Rachel is saying! The gaps that provided a bit of pacing and breathing space now feel like yawning chasms!

It may not look pretty, but it does still sound good, which is why this is often referred to as the radio edit; it's effectively an edit that is focused on the audio, not the visual. When creating a radio edit, you want to remove any large or small bits of audio that may detract from the message.

> **TIP**   To get a feel for how well a radio edit works, try closing your eyes during playback to just listen to the spoken words. If you can't hear any obvious edits, and the pacing of the speech sounds natural, that's generally the key to a good radio edit.

11   Re-enable the clips on the V2 track by clicking the Enable Video Track button in the timeline track header.

Using the Smooth Cut Transition

Now that you've seen how the edit is currently shaping up, and how the B-roll clips have been used to cover the different jump cuts in the interview, it's time to put the finishing touches on this timeline by adding a few more sound bites from the interview and painting them with appropriate B-roll.

First, though, you will fix a nasty jump cut at the start of the edit.

1   Click the Index button in the top left of the interface.

The Edit Index window opens below the media pool.

2   Click the Markers tab in the Index window to display a list of the markers in the current timeline.

3   Double-click the thumbnail image for the first marker named "Jump Cut" to move the timeline playhead to the first marker, at the edit between the first and second interview clips.

> **TIP** You can filter which colored markers are listed in the Markers Index using the Options menu (three dots) at the top right of the window.

4   Click the Detail Zoom button and press / (slash) to preview the cut.

This is a classic jump cut that distracts from the story about the merits of this vegan restaurant, but the director still wants to see as much of Rachel as possible before having to resort to using the B-roll footage as cutaways.

5   Open the Effects Library, select Video Transitions and, in the Dissolve category, locate Smooth Cut.

> **NOTE** With the Index window open, the Effects Library opens in place of the media pool.

6   Double-click the Smooth Cut transition to add it to the edit point.

7   Trim the Smooth Cut transition in the timeline to 4 frames (00:04 in the tooltip's gray numbers).

> **TIP** When trimming a transition, the gray numbers in the tooltip refer to the adjustment you're making, whereas the white numbers refer to the transition's new total duration.

8   Click the Full Extent Zoom button and press / (slash) to play around the selected transition.

You should see that Resolve successfully manages to blend the two sides of the jump cut into what appears to be a single take!

While this might seem like you'll never have to worry about using cutaways on an interview ever again, it's worth remembering that you should use this transition with great care; you don't want to change or misrepresent the meaning of something your interviewee said through your editing, no matter how subtly!

Importing a "Selects" Timeline

Now you'll turn your attention to the rest of this edit, starting with the end of Rachel's interview. While you could sit and continue watching all the unused interview footage and slowly start to piece together the final part of the interview, it's common for the director, edit producer, or even edit assistant, to put together a selection of clips they would like you to use or that they think might be useful. Often, they might work on their own systems, reviewing footage before passing a "selects" timeline to you for the fine tuning and to incorporate into the main edit.

This is what has happened in this case. The director of the "Welcome to the Pantry" promo has a cloned copy of the same media files you're currently working with and has used the free version of DaVinci Resolve 18 to pull together the final part of the interview with Rachel, together with a few extra clips they'd like you to use as cutaways.

Rather than sending a copy of the whole project for you to open and find the correct version of the timeline, they have simply sent you the relevant timeline as a .drt file.

> **NOTE** The extension .drt refers to a DaVinci Resolve timeline file and is an efficient way of transferring timelines from one Resolve project, or Resolve system, to another. Similar to an .xml or .aaf, a .drt file is a binary file that contains only the relevant information needed to rebuild a specific timeline within a DaVinci Resolve project. DaVinci Resolve timeline files do not contain any media, so the recipient system would need a copy of the media to be able to play the imported timeline, but this also means that the files are very small and therefore can be easily transferred across a network, via cloud storage or even email. You can create a .drt file from your currently active timeline by choosing File > Export > Timeline or by pressing Shift-Command-O (macOS) or Shift-Ctrl-O (Windows) and choosing DaVinci Resolve Timeline Files (*.drt)" from the file type menu.

1   In the media pool, select the Timelines bin. This is where you will import the .drt file into your project.

2   Choose File > Import > Timeline or press Shift-Command-I (macOS) or Shift-Ctrl-I (Windows).

3   Navigate to R18 Editors Guide/Lessons/Lesson 03, select **RADIO EDIT START.drt**, and click Open.

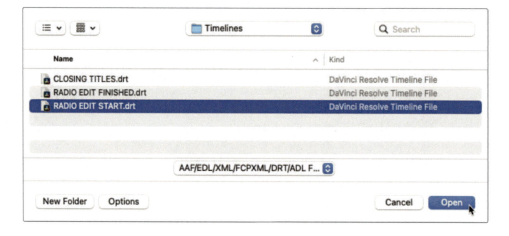

The imported timeline appears in the Timelines bin and automatically opens in the timeline window.

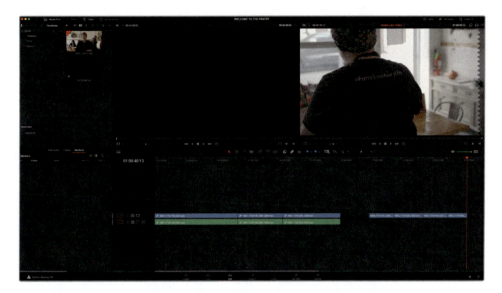

This timeline contains a basic assembly of three interview clips followed, after a short gap, by four additional B-roll clips. This is the footage that the director would like you to incorporate into the main edit. However, since the directors aren't skilled editors themselves, they have left much of the detailed work to you.

A common workflow for many editors is to use multiple different timelines to work on small sections of the overall edit; this could be the scenes of a feature film, the different parts of a broadcast TV show, or different sections of a corporate video. These scenes or parts often go through numerous iterations and changes themselves before being edited together into a final timeline that will be the version delivered to the distributor, broadcaster, or video streaming site.

Working Across Multiple Timelines

In the following steps, you'll use the timeline you've just imported to refine the new interview into a radio edit before incorporating it and the new B-roll into the main "Welcome to the Pantry" (WTTP) timeline.

1 In the **RADIO EDIT START** timeline, click the Full Extent Zoom button to see the whole timeline, and adjust the height of the audio tracks to reveal the clips' audio waveforms.

2 Play the first three interview clips to review the parts of the interview the director has chosen.

The first thing you'll probably notice is that Rachel's audio is only coming out of the left speaker. This is a common issue with people like the director putting together selects timelines like this because they are unaware of how Resolve is handling the audio in the different tracks. In this case, the director has simply placed the clips in the timeline without appreciating that the interview audio is mono, and the timeline track has defaulted to stereo. While this is not a huge problem in terms of this edit, it's still annoying.

3 Right-click the track header for A1 in the **RADIO EDIT START** timeline and choose Change Track Type To > Mono so the interview audio now plays out of both speakers correctly.

Whenever you're editing an interview such as this, or any piece of dialogue, it's useful to have a clear view of the clips' audio waveforms. To further enhance the height of the waveform, you can always adjust the clip's audio level. In this case, you can normalize the clip audio for a better view of the waveform. Don't worry; you can always normalize it again later if necessary.

4 Select all the interview clips in the timeline, and then right-click them and choose Normalize Audio Levels.

5 Using Sample Peak Program as the Normalization Mode, leave the Target Level set to -9 dBFS, but ensure that the Set Level is set to Independent, and then click Normalize.

> **TIP** You can always adjust track heights directly in the timeline by dragging the timeline track divider in the track headers or by using the Video and Audio Track Height sliders in the Timeline View Options menu.

6 Play the first clip again and stop playback when you hear the first time that the interviewee says, "Umm," around 2 seconds from the beginning.

7 Click the Detail Zoom button in the toolbar or hold Option (macOS) or Alt (Windows) and use the scroll wheel on your mouse, or scrolling function on your trackpad, to zoom in on the timeline.

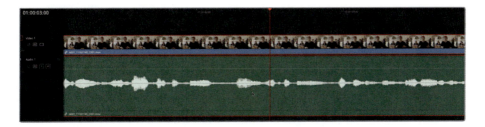

You should be able to clearly identify the problematic "Umm" in the waveforms.

8 Press the B key to switch to Blade Edit mode and click just before and just after the unwanted "Umm" to isolate the unwanted portion of the interview into a separate clip.

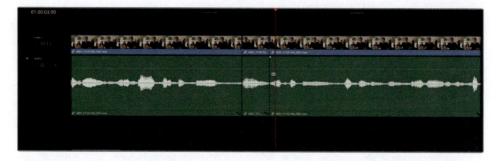

The dotted lines on the edit points represent through edits—that is, edits that are visible on the timeline but play back smoothly because no frames have been removed from either side of the cut. The edits were also added to both the audio and video parts of the clip because the timeline-linked selection was active.

> **TIP** To remove an unwanted through edit, in the timeline, place your playhead after the through edit and choose Timeline > Join Clips or press Option-\ (backslash) in macOS or Alt-\ (backslash) in Windows.

9 Press A to return to Selection mode.

10 Select the "Umm" clip, and press Shift-Delete (or Backspace) to perform a ripple delete.

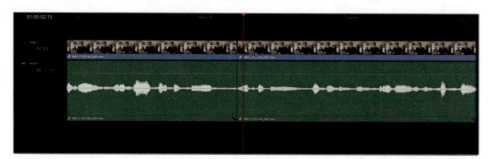

11 Place the playhead over the new edit point and press / (slash) to play around the edit.

You're aiming to have this audio edit be as inconspicuous as possible. When you play around the current edit, try listening to the cut without watching the picture. Does it sound as if an edit is there? If it does, you'll need to do a little finessing, which you'll look at soon.

12 Continue playing what is now the second interview clip until you hear the next problem where Rachel says, "We're taking dishes and flavors." She stutters and says the word "and" twice. It's a simple task to tidy this up.

13 Use the JKL keys to play forward and backward to position the playhead just before the first "and."

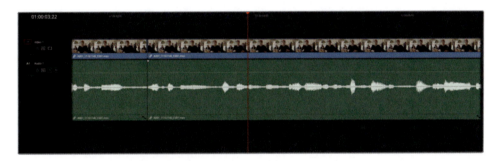

14 Press Command-B (macOS) or Ctrl-B (Windows) to cut the clip.

15 Jog forward until the playhead is located before she says the second "and." Press Command-B (macOS) or Ctrl-B (Windows) again to add a second through edit.

16 Move your playhead over the isolated "and" in the timeline and press Shift-V to select the clip under the playhead.

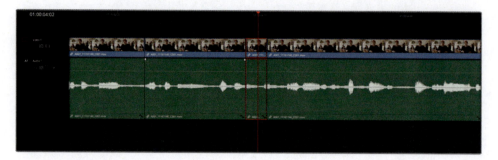

17 Press Shift-Delete (or Backspace) to ripple delete the clip.

18 Press / (slash) to play around the new edit.

Once more, try listening to the edit without looking at the jump cut you've just created. Again, although you're trying to create a natural-sounding interview that won't distract the audience, don't worry if the current edit is not as smooth as you would like; you'll finesse it soon.

19 Keep playing through the interview. The next portion you'll remove is the "Umm" just after she says "textures." This time, you'll place timeline In and Out points to remove this unwanted portion of the interview.

20 In the timeline, place your playhead at the start of the "Umm" and press I to add an In point.

21 Jog the playhead forward six or seven frames until you hear her start to say, "and really making them." Press O to set an Out point just before she says, "and."

You've now set In and Out points around the portion of the interview you want to remove.

22 Press Shift-Delete (or Backspace) to ripple delete the contents between the In and Out points in the timeline.

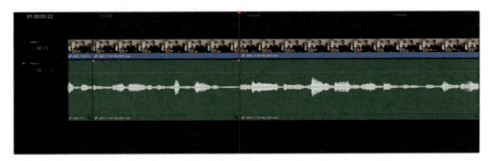

23 Press / (slash) to preview your new edit.

Well done. You have just removed three small portions of the interview using three slightly different techniques that have essentially achieved the same thing. As is often the case when editing, there is no simple right or wrong way of doing things, just more efficient or less efficient ways. As an editor, it's up to you to decide which tools and techniques suit your style and approach.

Extra Credit

The director has also asked if you could tidy up a specific part of the interview; she doesn't like the interviewee's use of the word "palatable" to describe the food. This is a slightly more subjective cut, but when trying to put together the best radio edit description of the food at the restaurant, it may be desirable to aim higher than simply "palatable." Do you think you'd be able to edit out the words "palatable and" so that she simply says, "making them delicious"?

Refining the Radio Edit

As you know, DaVinci Resolve has very comprehensive trimming options easily performed using the mouse. However, when you're making small, subtle, but very precise changes to an edit—often adding or removing single frames—it's useful to perform most of your trimming using the keyboard. In doing so, you'll exercise the most precise control over each of your edits. Learning how to get the best from Resolve's trimming options is an important step in choosing the best technique for any given situation.

You'll use keyboard shortcuts exclusively to select and trim the various cuts you've made in the previous steps.

1   Press Home to position the playhead at the start of the interview clips.

2   Press the Down Arrow to go to the first cut in that clip.

3   Press / (slash) to preview the edit.

Listen closely to the audio edit you created in the previous steps. Does it sound like a natural, continual part of her speech pattern? Does she fully pronounce all the words without having any part of what she's saying cut off? If so, can you identify where the problem lies by looking at the waveform?

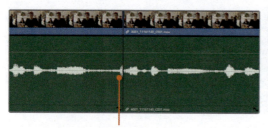 

Waveform indicates outgoing clip   Waveform indicates incoming clip
        might need trimming   might need trimming

Identifying how to adjust the edit effectively, and whether to add or remove frames from either the outgoing or incoming clips, is a skill that will come only with practice.

Don't worry if this all seems a little overwhelming right now. For the moment, you will simply learn to precisely adjust the cuts you made in the interview using the keyboard.

4   Press T to enter Trim mode.

5   Press V to select the edit point nearest the playhead.

6 Press the U key to toggle the selection until only the outgoing (left) side of the cut is selected for ripple trimming.

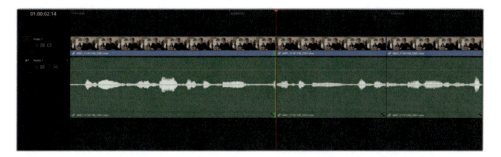

7 With the outgoing side of the edit selected, press , (comma) to trim the selected edit one frame to the left or . (period) to trim it one frame to the right, depending on whether you need to add or remove frames, respectively.

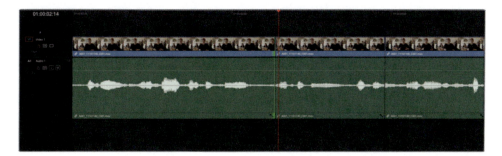

> **NOTE** In the example in the preceding images, one frame was trimmed from the end of the outgoing clip using , (comma) to remove the unwanted portion.

8 Press / (slash) to play around the selected edit to review the change.

9 If necessary, press U twice to change the trim to the head of the incoming shot.

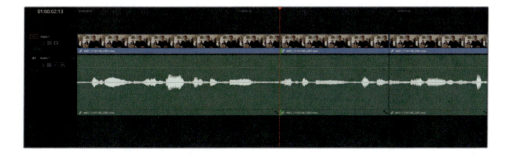

10 This time, because you have the incoming clip selected, press , (comma) to add a frame or . (period) to remove a frame as needed.

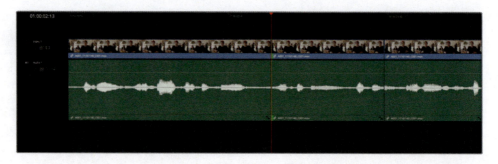

> **NOTE** In the example in the preceding images, one frame was added to the beginning of the incoming clip using , (comma).

11 Press / (slash) to play around the cut point and check your trim decision.

Continue to refine the edit until you are happy with the results. Then you can move on to the next cut.

12 Press the Down Arrow to go to the next cut in your timeline. This edit is selected in the direction you last selected the previous edit.

13 Press / (slash) to play around the cut point and to determine what you need to trim.

14 Decide whether the edit needs to be refined, and press U to toggle to the appropriate trimming operation.

15 Press the , (comma) or . (period) keys to refine the edit, adding or removing frames from either the tail of the outgoing clip or the head of the incoming clip.

16 Don't be afraid to press / (slash) to review your trim at any time, and continue to refine the edit if necessary.

When the audio edit sounds good, you can move to the next cut.

17 Press the Down Arrow to go to the next cut in that clip.

18 Press / (slash) to play around the cut point and identify what you need to trim.

19 Press U to toggle the edit direction and press , (comma) and/or . (period) to refine the edit as required, pressing / (slash) to review your trimming.

20 When you have finished, press A to return to the Selection mode.

You have refined a small part of this interview using various techniques to remove unwanted parts of what the subject says, and you've refined the rest into a succinct description of her business. When chipping away at a longer cut with the goal of making it shorter, these keyboard-oriented trim commands enable you to make tiny, precise trims as you see and hear the results, which can be invaluable!

Completing the Radio Edit

Continue to edit the two additional interview clips the director has included in this timeline.

— Edit the next clip so that Miss Rachel says, "...vegan food is starting to get the recognition it deserves."

— Edit the final interview clip so she says, "Because it seems like it's a diet or a lifestyle of cutting things out when, in reality, we're just doing things completely differently."

Now that the radio edit is complete, it's time to add the new interview clips into the main timeline and complete the edit.

Subframe Audio Editing

While video trimming is limited to a project's frame rate, audio is captured using tens of thousands of samples each second. DaVinci Resolve lets you edit audio at the subframe level in the edit page, which enables a much more detailed ability to trim that means you can isolate subtle syllables or words that are slurred and make the edits sound clean and clearer.

You don't need to do anything to activate this feature; it's how the edit page automatically works with audio. However, when trimming audio at the subframe level, it's best to have Snapping and Linked Selection turned off for the timeline. Zoom in as far as you can and do all your trimming with the mouse.

If you need greater precision than that offered by subframe audio editing, you'll need to utilize the editing functions on the Fairlight page, where audio can be trimmed precisely to the sample level.

Editing Between Timelines

Once you have a nice radio edit, you'll want to incorporate it into the much longer interview that's already been cut. DaVinci Resolve provides numerous ways in which you can edit footage from one timeline into another.

For consistency, a completed version of the radio edit timeline is available for you to import into this project to complete the rest of the steps in this lesson.

1   Select the Timelines bin and choose File > Import > Timeline and navigate to DR18 Editors Guide/Lessons/Lesson 03 /RADIO EDIT FINISHED.drt and click Open.

As before, the imported timeline is added to the selected bin and automatically opens in the timeline window.

2   Click the Full Extent Zoom button to see the entire timeline.

This is a version of the edited interview and the suggested B-roll clips, but with a short gap added to the interview for pacing.

3   Press Home to return the playhead to the beginning of this timeline and play the edited interview.

You now need to incorporate this edited interview into the main WTTP FINE CUT timeline.

DaVinci Resolve provides several ways of moving these edited clips from one timeline to another. For example, you could just copy the clips and paste them into the destination timeline. However, in the next steps you'll learn how you can use more sophisticated techniques to edit from one timeline into another.

To work across multiple timelines, you'll start by opening an existing timeline in the source viewer.

4 In the media pool, locate the WTTP FINE CUT timeline and drag it into the source viewer.

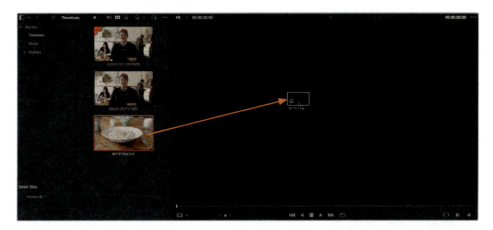

Scrubbing through the source viewer now scrubs through the WTTP FINE CUT timeline.

Now that you have the two timelines open—RADIO EDIT FINISHED in the timeline viewer and WTTP FINE CUT in the source viewer—you no longer have a need for the media pool, since the clips in one timeline will become the source for the other timeline.

You can now mark the clips you want to edit from the RADIO EDIT FINISHED timeline.

5 Select all the edited interview clips in the **RADIO EDIT FINISHED** timeline and press Shift-A to mark their total duration.

Currently, the **WTTP FINE CUT** timeline is the source for the **RADIO EDIT FINISHED** timeline, whereas you want them to be the other way around.

6 Choose Timeline > Swap Timeline and Source Viewer or press Command-Page Up (macOS) or Ctrl-Page Up (Windows).

The timeline in the source viewer—**WTTP FINE CUT**—is now open in the timeline viewer, and the timeline that was in the timeline viewer—**RADIO EDIT FINISHED**— is now the source, together with the In and Out points of the marked clips.

7 In the Markers Index, double-click the thumbnail for the second red marker, named "New Interview Here".

> **TIP** You can quickly jump between markers by choosing Playback > Next > Marker or Playback > Previous Marker, or by pressing Shift-Down Arrow for the next marker, or Shift-Up Arrow for the previous marker.

8 Ensure that the video and audio destination controls are set to V1 and A1, respectively, and make an Overwrite edit.

The clips are edited from the source timeline into the destination timeline as a regular clip would. However, the clips are currently displayed as single compound clip.

> **NOTE** You will learn more about compound clips and how you can work with them in Lesson 7.

9 Select the compound clip in the timeline and choose Clip > Decompose in Place.

The original source clips from the RADIO EDIT FINISHED are now displayed and can be edited as normal clips.

> **NOTE** These clips are now completely independent from the clips in the original source timeline, as if you had edited them into this timeline separately.

Decomposing Compound Clips on Edit

You can use the source timeline just like you would use any clip currently open in the source viewer. Next, you'll use the B-roll clips included in the **RADIO EDIT FINISHED** timeline as cutaways for the main edit.

1   In the **WTTP FINE CUT** timeline, move the playhead to the edit between the first and second clips you've just edited into this timeline.

2   Choose Timeline > Swap Timeline and Source Viewer or press Command-Page Up (macOS) or Ctrl-Page Up (Windows).

    The timelines switch viewers again.

3   Select the first two B-roll clips in the **RADIO EDIT FINISHED** timeline and press Shift-A to mark the selected clips.

4   Choose Timeline > Swap Timeline and Source Viewer or press Command-Page Up (macOS) or Ctrl-Page Up (Windows) to switch the timelines back.

5   Choose Edit > Decompose Compound Clips on Edit.

6   In the timeline, click the V2 destination control or press Option-2 (macOS) or Alt-2 (Windows) to target the V2 track.

7   Click the A1 destination control or press Command-Option-1 (macOS) or Ctrl-Alt-1 (Windows) to disable track targeting for the audio.

8 Make an Overwrite edit.

The clips from the source timeline are edited directly into the main timeline without the need to decompose the compound clip as a separate step.

You can also add In and Out points without having to open the source timeline.

9 In the timeline, press Down Arrow to jump to the next edit, select the next two clips, and press Shift-A to mark the selection.

10 Press Q to switch back to the source viewer and press the Down Arrow again until you see the shot of the two customers talking to Rachel.

11 Press I to place an In point at the first frame of this clip.

12 Make and Overwrite edit.

Just like a regular three-point edit, the clip from the timeline in the source viewer is edited into the fine cut timeline as a cutaway on V2.

Filling in the Atmos

Now, you just need to fill the gaps in the interview with a bit of atmos or wildtrack. Thankfully, since you already have sections of atmos used for this purpose on A2, it's simply a matter of copying those clips to the new gaps.

1 Select the third clip in the ATMOS audio track (A2) and press Command-C (macOS) or Ctrl-C (Windows) to copy the clip.

2 Place your playhead at the start of the gap just before the last red timeline marker.

3 Press Command-V (macOS) or Ctrl-V (Windows) to paste the copied audio clip.

4 Choose Playback > Next > Gap or press Option-Command-' (apostrophe) on macOS or Alt-Ctrl-' (apostrophe) on Windows to move to the next gap.

> **NOTE** The Next Gap and Previous Gap commands move the playhead to the appropriate gap on any auto select-enabled track.

5 Again, press Command-V (macOS) or Ctrl-V (Windows) to paste another copy of the clip.

6 Zoom in to the timeline and move and trim the pasted audio clips so that they cover the gaps in the interview audio.

7 When you've finished finessing the position of the new atmos clips, click the Full Extent Zoom button to zoom back out to view the whole timeline.

Working with Stacked Timelines

The preceding steps are a useful technique for editing the content of one timeline into another. However, it only allows you to see the content of one timeline at a time. Sometimes, though, you'll want to be able to see the content of both timelines simultaneously. Helpfully, the edit page timeline window can display multiple timelines in different tabs or stacked one above the other.

1 In the media pool, ensure that the Timelines bin is still selected and choose File > Import > Timeline or press Shift-Command-I (macOS) or Shift-Control-I (Windows).

2 If necessary, navigate to R18 Editors Guide/Lessons/Lesson 03/Timelines and select the CLOSING TITLES.drt file and click Open.

The CLOSING TITLES timeline is imported, added to the Timelines bin in the media pool, and opens in the timeline window.

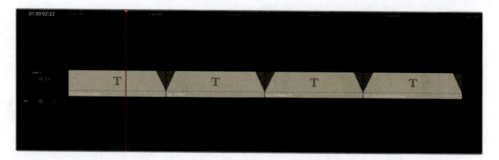

This timeline consists of four titles for the closing section of the "Welcome to the Pantry" promo. Each title has a fade at the top and tail (with the exception of the first title) and a dynamic zoom applied.

3 In the top left of the timeline toolbar, click the Timeline View Options menu and choose the first option to enable stacked timelines.

Tabs appear above the timeline, with the timeline's name. You can use the pop-up menu on this tab to access any other timeline from your project (similar to the Timeline Viewer pop-up menu) or open additional tabs using the plus button to display multiple timelines.

In addition to the tabs, another type of Add Timeline button appears in the top-right corner of the timeline window.

4 In the upper-right corner of the timeline, click the New Stacked Timeline button.

A second timeline window appears below your currently active timeline.

5 If necessary, resize the clips in the upper timeline window so you can still see the overall structure of the clips.

6 In the upper timeline, close the tab for the **CLOSING TITLES** timeline.

7 In the lower timeline tab, click the Select Timeline pop-up menu and choose the **CLOSING TITLES** timeline.

The **CLOSING TITLES** timeline reopens in the lower timeline window.

8   Take a moment to adjust the track heights of the upper timeline so you can see the contents of the **WTTP FINE CUT** timeline clearly.

9   Select all the title clips in the lower timeline and drag them to the end of the last interview clip in the upper timeline.

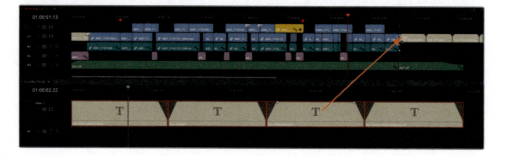

The clips are instantly copied from the **CLOSING TITLES** timeline into the main **WTTP FINE CUT** timeline.

Now that you have edited the titles into your main timeline, you can close all the timeline tabs you no longer need.

10 With the **WTTP FINE CUT** timeline the currently active timeline (as denoted by the red timeline tab), click the Timeline View Options menu and click the Stacked Timelines option to turn the stacked timelines off and return to a single timeline view.

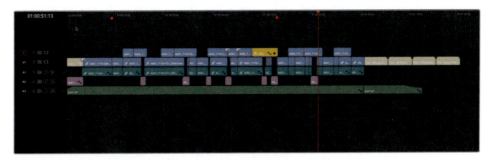

Now all you need to do is trim the new titles to the correct duration.

11 Using Selection mode, select the four title clips you've just added to this timeline.

12 With the clips selected, press T to switch to Trim Edit mode.

13 Choose Clip > Change Clip Duration or press Command-D (macOS) or Ctrl-D (Windows).

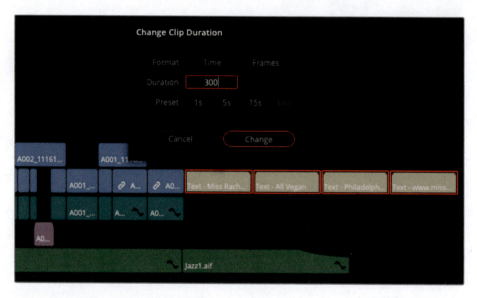

14 In the Change Clip Duration window, type **300** in the Duration field and click Change.

All the clips are instantly trimmed to a duration of 3 seconds and, because you were in Trim Edit mode when applying the change, have rippled the timeline so no gaps remain.

15 Return the playhead to the beginning of the timeline and review the final, completed edit for the "Welcome to the Pantry" promo.

> **NOTE** You can import a finished version of the WTTP FINE CUT timeline by choosing File > Import > Timeline and navigating to R18 Editors Guide/ Lessons/Lesson 03, selecting WTTP FINE CUT FINISHED.drt, and clicking Open.

Many complex edits are often broken down into different stages across multiple timelines, each of which might go through several different versions before being assembled into the finished timeline for delivery. By carefully managing each of these timelines, you'll find that it's a more efficient way of working through an edit than trying to do everything in one all-encompassing timeline.

Lesson Review

1   What is known as a "radio edit"?

    a)   When your edit will be used on radio as well as television

    b)   A cut-down version of your film used by reviewers

    c)   A technique of concentrating on the audio before editing the visuals

2   Which transition could be used to help make jump cuts less noticeable?

    a)   Smooth wipe

    b)   Smooth dissolve

    c)   Smooth cut

3   Where can you find the option to show stacked timelines?

    a)   Drag one timeline on top of another to display two timeline windows.

    b)   The Timeline View Options menu

    c)   Clip > Decompose in Place

4   How can you open a timeline in the source viewer?

    a)   Double-click the timeline in the media pool.

    b)   Right-click the timeline in the media pool and choose Open in Source.

    c)   Drag the timeline from the media pool to the source viewer.

5   Which option stops DaVinci Resolve from automatically editing the content from one timeline to another as a compound clip?

    a)   Clip > Decompose in Place

    b)   Timeline > Decompose Compound Clips On Edit

    c)   You cannot stop DaVinci Resolve from editing clips from one timeline to another as a compound clip.

Answers

1   c) A "radio edit" is created when you focus on editing the audio of an interview to make it sound smooth and natural before turning your attention to the covering visuals.

2   c) Smooth cut is a special-purpose transition designed to make short jump cuts less noticeable by using optical flow processing to automatically morph a subject between two positions across the duration of the transition.

3   b) The Timeline View Options menu

4   c) Drag the timeline from the media pool to the source viewer.

5   b) Timeline > Decompose Compound Clips On Edit

Cutting a Dramatic Scene

Editing a dramatic scene is often done by establishing the location and cutting between shots as they would play out in real time. Commonly known as *continuity editing*, this technique is centered around cutting between two or more shots, alternating back and forth between each character as their dialogue and reactions warrant. In this lesson, you'll apply this continuity technique to a simple scripted scene. You'll start with one of the most firmly established conventions in cinema— the shot/reverse-shot—and see how the editing and trimming tools in DaVinci Resolve 18 can speed up this classic editing style.

In this lesson, you'll edit a short dialogue scene between two characters from the short sci-fi film Sync about human-like robots going rogue—written, produced, and directed by Hasraf "Haz" Dulull.

Time

This lesson takes approximately 60 minutes to complete.

Goals

Working with Separate Takes

Although a project like this is often scripted, rehearsed, and shot under controlled conditions, there are still a number of creative choices that need to be made when editing.

Watching each take and choosing the parts that feature the best performances is often the most time-consuming part of your entire editing process, but it is also a critical step in becoming familiar with the available content and identifying which shots and performances might or might not work. Often, you might need to work with different parts of different takes to get the best performances.

You'll start by importing the project for this lesson and relinking the necessary media files.

1   Open DaVinci Resolve and, in the Project Manager, click the Import button.

2   Navigate to R18 Editors Guide/Lesson 04, select the **SYNC SCENE.drp** project file, and click Open.

3   Once the project has been imported, double-click to open it and, if necessary, select the edit page.

4   In the media pool, click the Relink Media button and relink the media files.

5   Choose Workspace > Reset UI Layout.

6   In the timeline, click the Full Extent Zoom button and resize the timeline tracks so you can comfortably see all the clips in the timeline window.

The project is organized in a series of bins for either Agent Jenkins or Doctor Kominsky, referred to as AGENT J or DR K, respectively.

7 In the timeline, return the playhead to the start of the timeline and play through to review the scene.

The first shot in the timeline shows Dr K walking off stage after speaking at a conference of cybersecurity experts, shown in a previous scene. The main part of the timeline features the whole scene as shot from Agent J's camera angle; this is often referred to as the master shot because it lays out the structure of the scene. It is, however, missing the reverse shots of Dr K that not only show her speaking her lines but, crucially, also show the interaction between the two characters and other elements in this scene.

> **NOTE** Although it may appear that this scene has been shot across multiple cameras, it is not a multicamera shoot. Instead, the scene was shot single camera—a typical filming technique that requires the scene being run multiple times and filmed from different camera angles. The takes from these different camera angles are then edited together in a seamless scene, provided that continuity is maintained of course. Shooting single camera is a cost-effective method of production since it requires the minimum amount of crew and equipment, but it requires much more time on set since each scene needs to be shot multiple times from all the different angles required by the director. Similarly, the edit needs to be pieced together carefully and, because each angle and take is highly unlikely to have exactly the same pacing, multicamera editing techniques will not be appropriate. You'll learn more about how to edit multicamera productions in Lesson 5, "Multicamera Editing."

8 Select the DR K CAM bin.

This bin contains the various takes featuring Doctor Kominsky.

9   Select all three clips in this bin and drag them into the source viewer.

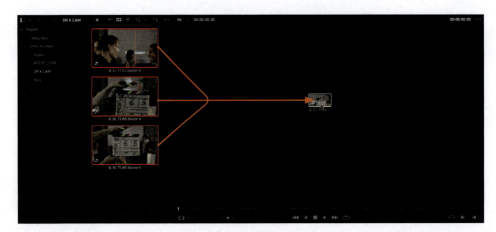

By dragging these clips into the source viewer at the same time, you can access them using the Recent Clips dropdown menu at the top of the viewer, providing an easy way to switch between the last 10 clips you loaded in the source viewer without having to locate them in the media pool.

10 In the source viewer options menu, select Show Full Clip Audio Waveform to help you see where the dialogue starts and stops.

11 Select **B_S6_T5 Doctor K WS** in the Recent Clips menu to load it into the source viewer.

12 Play the clip to review it.

This is the first of three takes of the scene from Doctor Kominsky's camera angle.

> **NOTE** The naming convention for the clips in this project are based on the camera setup ("A" for Agent Jenkin's angle and "B" for Doctor Kominsky's angle), the slate or shot number (e.g., S5), the take number (e.g., T1), a shot description (e.g., "WS" for wide shot or "CU" for close up), and finally which character the clip features—e.g., B_S6_T5 WS DOCTOR K. You'll learn more about how to organize and name your clips in Lesson 6, "Project Organization."

Editing for Continuity

You will use the single take in the timeline as a reference for editing dialogue and action from other takes, and you'll use some editing functionality that is unique to DaVinci Resolve to add close-ups and reaction shots that build both time and space continuity.

The chosen master shot focuses on the FBI agents (though it could just as easily have been Dr K's angle), so it's up to you to add reverse-angle shots of the doctor. As with the initial steps in the previous lesson, you are focusing on the pacing of the dialogue from these different takes in an attempt to hide the artifice of filmmaking and sell the idea of these characters having a seamless conversation shown from different camera angles. You will refine this rough cut further in later steps.

1 If necessary, play the timeline again to remind yourself of the scripted dialogue between these characters. Remember that the more familiar you are with the footage, the better you will be able to assess how each take will cut together and the more creative your choices are likely to be.

   Currently, after the doctor has left the stage to applause, the scene "opens" on an empty shot before the FBI agents walk in. You'll start by showing the doctor being interrupted by the FBI agents.

2 In the timeline, place the playhead at the start of the clip **A_S8_T2 WS Agent J** and mark an In point.

3 Play the clip in the timeline and mark an Out point just before Agent Jenkins's line, "I'm Agent Jenkins...."

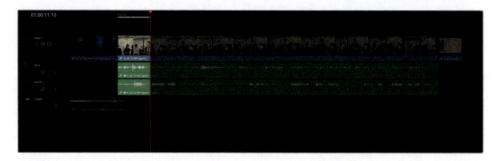

> **TIP** You can use the waveforms in the timeline to see where each of the characters starts and stops speaking, with Dr K on the A1 track and Agent J on the A2 track.

4 Press Q to switch to the source viewer and clip **B_S6_T5 WS Doctor K**.

5 Play the clip until after the man has walked away from Dr K, just as the woman is approaching her (at about 01:02:55:00).

6 Mark an Out point just before Agent J delivers his line, "I'm Agent Jenkins...."

Hold on. You have now managed to add a total of four points: an In and Out point in the timeline and another In and Out point in the source viewer. Also, while all four marked points are well placed, the resulting durations of your two selections differ: in the upper-left corner of both the source and timeline viewers, you have the current marked duration—around 7 seconds for the source and just short of 5 seconds for the timeline (depending on where you actually placed your timeline Out point).

As you are now no doubt aware, many of the editing functions in DaVinci Resolve follow the standard rules of three-point editing; therefore, only three of these four points would be valid for a standard Overwrite or Insert edit. You can, however, preview which marks Resolve will use.

7   Choose View > Show Preview Marks.

The preview marks display a blue indicator in the source viewer to highlight the shorter duration marked in the timeline, meaning the Out point in the source viewer will not be honored in a standard three-point edit.

8   Press F10 or click the Overwrite Clip button above the timeline and preview the edit in the timeline.

Performing an Overwrite limits the edit based on the shorter duration marked in the timeline. This leaves a jarring cut from the Doctor K shot to Agent Jenkins introducing himself. If you were to proceed with this edit, you would have to trim the outgoing clip of the doctor to correct for the continuity of the scene and recover the missing dialogue from Agent Jenkins.

Fortunately, Resolve provides a unique edit that helps when editing scenes such as this.

9   Choose Edit > Undo or press Command-Z (macOS) or Ctrl-Z (Windows).

10 Drag the clip in the source viewer to the timeline viewer and over the Ripple Overwrite option in the edit overlays.

Ripple Overwrite overwrites the marked duration of the clip in the source viewer to the marked duration in the timeline, rippling the timeline automatically in order to take into the account the differences in the two marked durations. And, yes, Ripple Overwrite has used all the In and Out points you have marked, which means it is a *four-point edit*.

In the timeline, play over the edit to verify that you have successfully overwritten the unwanted part of the clip with a shot that introduces the doctor. This time, the frames that were previously ignored in the source viewer by the usual Overwrite edit have now been included by the Ripple Overwrite edit, with the timeline becoming a few seconds longer as a result.

Using Ripple Overwrite, you can continue editing this scene.

11 Play the timeline and add an In point after Agent Jenkins says, "We need you to come with us right away," and an Out point as Agent Jenkins turns away, just before Dr Kominsky says, "Yeah, yeah"—a duration of just over 3 seconds.

12 Press Q to switch back to the source viewer and play clip **B_S6_T5 WS Doctor K**, adding an In point just after Agent Jenkins's line, "We need you to come with us right away," and an Out point as Doctor Kominsky turns to her associate and before she says, "Yeah, yeah. I'm sure it's fine."

13 Make a Ripple Overwrite edit using the timeline viewer overlays or by pressing Shift-F10.

14 Continue playing the timeline, setting the next In point just before Doctor K mutters, "OK," and the Out point just before Agent J says, "We need your help."

15 Press Q to switch back to the source viewer and set an In point after the doctor's associate starts to walk away and an Out point just before Agent Jenkins says, "We need your help...."

16 Make a Ripple Overwrite edit.

As you can see, by using the unique Ripple Overwrite edit, you can quickly block out the scene, using the dialogue as a guide to the pacing. Of course, there's more to consider, but at the moment you're only interested in getting a rough cut together in this timeline.

Using Alternate Takes

1 Continue playing the timeline, setting the next In point just as Doctor K says, "My help?" and the Out point just as Agent K turns back and before he says, "OK, we don't have a lot of time...."

2 Press Q to switch back to the source viewer and play back to listen to the equivalent line from the doctor's camera angle.

Ouch! That's awful audio on this take! Obviously, the actor playing Doctor Kominsky is wearing a microphone, and as she grabs the lanyard, she also catches the mic. Obviously, this makes the take unusable and is one of the many reasons why additional takes are recorded!

3 From the source viewer Recent Clips menu, choose **B_S6_T3 WS Doctor K**.

4 Place the playhead about three-fifths of the way through the clip and find the line you're looking for, when Doctor K says, "My help? You'll have to do better than that." (at about 01:05:45:00).

5 Add an In point just before she says, "My help" and an Out point just before Agent J says, "OK, we don't have a lot of time here…."

You might still hear the actor catching the microphone, but it's not nearly as bad as the previous take. It's also isolated from her actual dialogue, so it should be easily edited out later.

6 Perform a Ripple Overwrite edit.

7 Continue playing the timeline and mark an In point just after Agent J says, "…deadly, weaponized package," and an Out point just before he says his next line, "Well, it's possible Doctor Kominsky…."

8 Press Q to switch back to the source viewer and select B_S6_T5 WS Doctor K from the Recent Clips list.

9 Add an In point just after Agent J says, "…deadly weaponized package."

> **NOTE** You might want to take note of Doctor Kominsky's reaction to this line. It's a nice shake of the head in disbelief that you might want to include.

10 Set an Out point in the source viewer just before Agent J says, "Oh, it's possible…."

11 Make a Ripple Overwrite edit.

The final edit you will make will be a standard reaction shot from Doctor Kominsky.

12 Play the timeline and set In and Out points around Agent J's line, "We need you to tell us everything you know about the Sync: the routes, their drop points…."

13 Press Q to switch to the source viewer, play forward, and set an Out point after Agent J says, "...their drop points".

14 Choose Mark > Clear In or press Option-I (macOS) or Alt-I (Windows) to remove the In point.

Removing the In point in the source will enable you to backtime the reaction shot when you edit it into the timeline.

15 In the timeline, click the track destination controls for A1 and A2 to disable audio from the source clip.

16 Make an Overwrite edit.

Fantastic! With the help of the Ripple Overwrite function, you have been able to quickly put a rough cut of this scene together. As always, though, this is only the first step when editing; next, you'll learn how to work with evaluating different takes before polishing the scene in Trim Edit mode.

> **NOTE** If you need to catch up before moving to the next step, select the TIMELINES bin and choose File > Import > Timeline, navigate to R18 Editors Guide/Lesson 04/Timelines and select SYNC SCENE ROUGH CUT CATCHUP 1.drt and click Open.

Comparing Different Takes

When cutting dialogue, you can easily fall into the trap of cutting based solely on words. But dialogue editing is trickier than that. You must pay attention not only to the words but also the eyes, the mouths, and the body language of the performers. All these performance elements are essential to establishing the emotional pacing of a scene and establishing the scene's storyline. So, even though your cut may maintain dialogue continuity, you might want to search out alternate takes that feature superior performance elements, alternative looks, or body or camera movements.

Comparing different takes often means repetitively revising your timeline. For example, imagine that the director wants to see what a wide shot of the doctor would look like instead of the close-up. Sounds simple enough to edit one take in place of another, doesn't it? However, most editors will tell you that a director is rarely satisfied with that and often wants to see multiple choices in quick succession or switch back and forth between alternate takes. DaVinci Resolve can make this process a lot easier with the Take Selector.

Before you start this process, though, it would be prudent to duplicate the current timeline.

1 Select the TIMELINES bin and right-click the SYNC SCENE ROUGH CUT timeline and choose Duplicate Timeline to create a copy of the timeline.

2 Return to the current timeline and position the playhead at the start of the fourth clip in your timeline, the second instance of B_S6_T5 WS Doctor K.

The director wonders aloud how this wide shot compares with an alternate close-up take that was shot.

3 From the source viewer clips list, select B_S5_T1 CU Doctor K to open the close-up take in the source viewer.

4 Mark an In point just before Doctor K says "Ummm" (at around 01:01:00:00) and an Out point just before she says "yeah, yeah…" (about 2 or 3 seconds later).

To audition this take in the timeline, without actually replacing the current take, you will enable Resolve's Take Selector.

5 In the timeline, right-click **B_S6_T5 WS Doctor K** where your playhead is and choose Take Selector or select the clip and choose Clip > Take Selector.

The Take Selector is activated on the selected timeline clip.

The Take Selector acts as a container for multiple clips. While you'll see only one of those clips when you play the timeline, you can switch between the clips at any time.

6   From the source viewer, drag **B_S5_T1 CU Doctor K** anywhere into the timeline to automatically add it to the activated Take Selector.

TIP You can add as many takes as you need to the Take Selector.

The Take Selector now shows the two clips stacked on top of each other.

The most recently added take is the active take, but you can change that by clicking the clip you want to view in the timeline. However, you'll notice that the close-up take you just added has a slightly different duration than that used for the wider take.

7 In the upper-right corner of the Take Selector, click the Ripple button.

The timeline adjusts to fit the shorter take's duration.

8 Play the timeline to review the alternative take.

The director decides that the close-up is a better reaction shot to use at this point, so you'll make the new take a permanent part of the timeline.

9 In the Take Selector, click the upper clip to choose that take.

10 In the upper-left corner of the Take Selector, click the Close button to collapse the Take Selector stack.

> **TIP** At this point, you can always reopen the Take Selector and continue reviewing the different takes for this opening shot by double-clicking the Take Selector icon in the lower-left corner of the clip in the timeline.

11 Right-click the clip and choose Finalize Take.

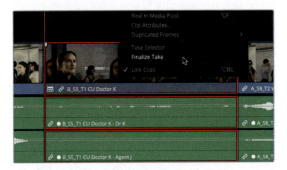

Finalizing the take removes the Take Selector and all the alternate takes from the timeline, leaving the director's final choice as a stand-alone clip.

Tidying Up the Audio

Before you turn your attention to finessing this scene, there's now an opportunity to tidy up the audio for the dialogue clips. This will help you eliminate any unwanted or unneeded parts of the audio clips, leaving you with just the words spoken by the actors.

1   In the timeline, ensure that Snapping and Linked Selection are disabled. This will make it easier when selecting and trimming the audio clips.

2   Ensure that you are in Selection mode and trim the start of the first audio clip on A2 forward by about 5 seconds (+05:00 in the tooltip) so it starts when Agent J says his first line.

3   Trim the end of the same clip backward by about a second to the end of Agent J's line.

4   On the next clip, select the clip on A1 and press Delete (Backspace) to remove it.

This leaves just Agent J's dialogue track for this clip.

The next two clips are for dialogue spoken by Dr K, even though you see Agent J for one of them.

5   Select the next three audio clips on A2 and press Delete (Backspace) to remove them from the timeline.

6   Continue working through the rest of the timeline, removing the unneeded audio clips on A1 when Agent J is speaking, and the clips on A2 when Dr K is speaking.

When you've finished, you should have a timeline that features only the audio that you need, with each character on their own separate track. This will also make it easier when trimming the different clips in the timeline, since you don't have to worry about clips that are not needed. You'll complete the audio mix for this scene in Lesson 8, "Editing and Mixing Audio."

> **NOTE** If you need to catch up before moving to the next step, select the **TIMELINES** bin and choose File > Import > Timeline, navigate to R18 Editors Guide/Lesson 04/Timelines and select SYNC SCENE ROUGH CUT CATCHUP 2.drt and click Open.

Utilizing Split Edits

A straight cut, with which audio and video start and end simultaneously, can be quite abrupt and a little jarring. A *split edit*, often referred to as an *L-cut* or *J-cut* because of the implied shape it creates in the edit, delays either the audio or video cut of a clip. Staggering the cuts in this way can create a more natural rhythm between shots.

The most common split edit is the J-cut, with which you first introduce the sound of the next shot and then cut to the picture slightly later. This is the way that most of us perceive the world around us. For example, when you hear a car horn in the street, you look for the source of that sound a fraction of a second later.

An L-cut leads with the picture edit before the audio edit and is often used when you want to show a character's reaction to something happening or being said, before their response.

Resolve provides multiple ways to create J-cuts and L-cuts. You'll start by creating a J-cut in which you hear upcoming clip audio first before letting the picture cut.

Extend Edit

Extend edit is a common way to create a split edit, since it automatically trims a selected edit point to the playhead position.

1   Ensure that both Snapping and Linked Selection are still disabled in the timeline toolbar, and then place the timeline playhead between the second and third clips and click the Detailed Zoom button, adjusting the zoom as appropriate so you have a comfortable view of the clips in the timeline.

2   Press the / (slash) key to review this edit.

As the FBI agent introduces himself, it might be better to hold on to the shot of the doctor so you can see her reaction. Also, hearing Agent J's voice is a good motivation for the cut to the next shot since we'll want to see them in vision once they start talking.

3   Select the edit point between the video clips **B_S6_T5 WS Doctor K** and **A_S8_T2 WS Agent J**.

4   Now, move the playhead to the frame where Agent J begins to turn his head to introduce his partner.

5 Choose Trim > Extend Edit or press E.

The extend edit trims the selected video edit point to the position of your playhead in the timeline. Because you had both sides of the edit selected, this resulted in a roll trim and created a split edit. Press / (slash) to review the new split edit.

That edit now feels very natural in that you see the doctor react to the FBI agent's introduction, while the FBI agent's line of dialogue motivates the picture cut a second or so later. However, the edit now feels like it could be a bit tighter.

6 Press T to activate Trim Edit mode.

7 Press U twice to toggle the selected edit to an outgoing ripple edit.

8 Using , (comma), remove about 10 frames from the outgoing edit, reducing the amount of time between Doctor K's line and Agent J's response.

9 Press / (slash) to review the changes, making sure you're happy with the pacing between the two shots.

> **TIP** Don't forget that you can always add frames back on to this clip by pressing . (period).

When you're satisfied with this edit, you can turn your attention to the next edit.

10 Press the Down Arrow key to move to the next edit.

11 Press the / (slash) key to review the edit.

Now you'll trim this edit so that you see Doctor Kominsky's reaction sooner.

12 Select the video edit and position the playhead just as Agent J says, "Come with us."

13 Press E to perform the extend edit.

Again, the extend edit moves the selected edit point to the position of your playhead in the timeline.

14 Press / (slash) to review the new split edit.

Now, you see much more of Doctor K's reaction to the news. You can make much more of this moment by slowing down the pace.

15 Press U to toggle the selection to an incoming ripple edit.

16 In the timeline viewer timecode field, type **-12** (-00:12) and press Enter (Return).

The selected edit point ripples back by half a second, giving Doctor K that extra beat to realize what Agent J is telling her.

17 Press / (slash) to review the change.

Creating simple split edits like this is a quick and straightforward editing technique that is used all the time, not just in dialogue scenes such as this but in editing across all genres.

Dynamic Trimming

Another mode you can use in conjunction with either the Selection or Trim Edit modes is the Dynamic Trim mode. Many editors like to use dynamic trimming because it allows you to trim in real time as you are playing back. It takes a bit of practice to use effectively, but it can be a very fast way of making even precise adjustments to a timeline.

1 In the current timeline, click away from any clips to deselect all clips and/or edit points.

2 Place the playhead over the next edit: the cut between B_S5_T1 CU Doctor K and A_S8_T2 WS Agent J.

3 Press / (slash) to review this edit.

The pacing feels a little too tight, and you would like to see more of Agent J's exasperated reaction to the doctor's flustered reaction.

4 Press T to ensure that you are in Trim Edit mode.

5 In the timeline toolbar, click the Dynamic Trim mode, or press W.

> **NOTE** Dynamic Trim can be enabled when you are in Trim Edit or Selection mode.

When Dynamic Trim mode is enabled, several things happen. First, the edit point nearest the playhead is automatically selected, and the timeline playhead and Dynamic Trim button turn yellow to indicate that you are now in Dynamic Trim mode.

However, other changes are not as obvious. Your familiar keyboard shortcuts have now also changed their functions! If you've been using Spacebar to start/stop playback, this now becomes the equivalent to the Play Around Current Frame function, while the J and L keys now adjust the edit point in real time!

6   Press U to toggle to a rolling selection, with both sides of the video edit being selected.

7   Press and hold the K then J keys together to begin trimming slowly backward until before Agent J turns his head.

8 Press Spacebar to review the edit and adjust as necessary by holding K and L or J, depending on whether you want to trim forward or backward.

> **TIP** If you want to trim a frame at a time, the familiar , (comma) and . (period) shortcuts will continue to work. Alternatively, you can hold down K and tap J or L.

When you're happy with the edit, press Down Arrow twice to jump to the edit point between A_S8_T2 WS Agent J and B_S6_T5 WS Doctor K.

The audio and video edit points are automatically selected.

9 Click the video edit to select it separately from the audio edit or press Option-U (macOS) or Alt-U (Windows).

10 Press L to roll the edit forward as Agent J watches Doctor K's associate walk away and his eyeline returns to look at the doctor.

11 Press Down Arrow to jump to the next edit point.

12 Press Option-U (macOS) or Alt-U (Windows) to toggle the selection to just the video edit.

13 Press K and L to roll the edit forward slightly until Agent J says, "…matter of national security" and begins to step away in the incoming clip.

14 Press Down Arrow to select the next edit between **A_S8_T2 WS Agent J** and **B_S6_T3 WS Doctor K** and press Option-U (macOS) or Alt-U (Windows) to select the video edit.

15 Press L to roll the edit forward as Doctor K says, "My help?" and Agent J begins to step away from her.

TIP All the trimming functions you're familiar with, including slip and slide, are available to you in Dynamic Trim mode. To slip or slide a clip using dynamic trim, select the clip and start trimming using the JKL keys. To switch between slip and slide edits, press S or right-click the Dynamic Trim button in the timeline toolbar.

Putting Yourself to the Test

Continue to finesse this edit using any of the techniques you prefer to adjust the pacing and cuts between each shot. Remember, there are really no right or wrong ways to cut a scene like this, but these techniques have been used by editors around the world for decades, so they are a good place to start.

NOTE If you need to catch up before moving to the next step, select the TIMELINES bin and choose File > Import > Timeline, navigate to R18 Editors Guide/Lesson 04/Timelines and select SYNC SCENE ROUGH CUT CATCHUP 3.drt and click Open.

Filling the "Dead Air"

Once you've finished finessing the pacing of this scene, you'll probably have short gaps between some of the audio clips, particularly if you added pauses between the reactions and the next line of dialogue. To help fill this "dead air," you'll add an audio clip of some general background noise. This atmos, or wildtrack, could have been recorded on location at the time of the shoot, or it could be a piece of audio from a sound effect library.

1   Press N to enable snapping in the timeline.

2   Return the timeline playhead to the first shot of Doctor K talking to her associate.

3   From the Audio bin, select the clip **Atmosphere wild sound** and open it in the source viewer and add an In point near the start.

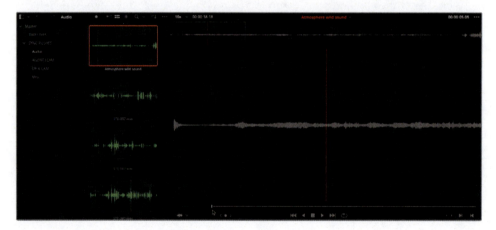

4   Ensure that the audio destination control in the timeline is active next to the DR K audio track.

> **NOTE** If the audio track selection control is disabled, Resolve won't let you edit this audio-only clip into the timeline in the next step.

5 Drag the audio clip to below A3 in the timeline.

The clip is automatically edited onto a new A4 audio track, but it's too short to reach the end of the scene.

6 Drag another instance of the same audio clip to this new track, starting about halfway through the scene.

7 Trim the end of the second version of the atmos clip to just before the final shot in the timeline.

8 Using the fade handles, add a short fade to the start of the first clip and the end of the second clip on track A4.

9 Return your playhead to the beginning of the timeline and play back the scene to review your hard work.

> **NOTE** If you need to catch up before moving to the next step, select the TIMELINES bin and choose File > Import > Timeline, navigate to R18 Editors Guide/Lesson 04/Timelines and select SYNC SCENE ROUGH CUT CATCHUP FINISHED.drt and click Open.

Well done! You have managed to edit a short dramatic scene between two characters using a variety of editing techniques, including Ripple Overwrite, extend edits, split edits, and dynamic trimming.

While this lesson has ostensibly been about editing drama, its underlying theme is continuity. Continuity editing involves matching screen direction, position, and temporal relations from shot to shot in support of the story. Hopefully, you now have a deeper understanding of how the editing tools in DaVinci Resolve can support this important underlying principle of editing.

Lesson Review

1   What do the preview marks on the timeline indicate?

   a)   They show where you can add markers.

   b)   They help determine where clips will be placed when making three-point edits.

   c)   They show where text and graphics will align on the timeline viewer.

2   True or false? Ripple Overwrite is a three-point edit.

3   True or false? All clips in the Take Selector should have the same duration.

4   Which of the following are commonly types of split edits?

   a)   J-cuts

   b)   K-cuts

   c)   L-cuts

5   True or false? Dynamic Trimming can only be used when the timeline is in Trim Edit mode.

Answers

1   b) Preview marks help you determine where clips will be placed whenever you execute a three-point edit.

2   False. Ripple Overwrite is a four-point edit that you use when the duration of the marked source clip is different from the duration marked in the timeline, and when you want the timeline to ripple to accommodate the difference in duration.

3   False. The Take Selector can contain clips of different durations. When switching between these clips, you can enable the Ripple Take button in the upper-right corner (to the left of the trash can button).

4   a) and c) J-cuts are when audio edits precede video edits, and L-cuts are when video edits precede audio edits.

5   False. Dynamic Trimming can be used when the timeline is in either Selection or Trim Edit mode.

Multicamera Editing

Multiple cameras running simultaneously are used for many types of productions, including music videos and reality TV programs, with even simple interviews benefitting from being shot on more than one camera.

The multicamera editing workflow in the edit page allows you to initially synchronize multiple clips and then easily cut between different camera angles without any further concern about sync issues. Utilizing the power of metadata provides even more flexibility when it comes to naming and ordering camera angles.

In this lesson, you'll explore the power of multicamera functionality in the edit page across two projects and learn how to solve some common challenges.

Time

This lesson takes approximately 60 minutes to complete.

Goals

Editing a Multicamera Interview

Accurately establishing the sync relationship between multiple camera angles is critical for a successful multicamera edit in the edit page.

1 Open Resolve, and in the Project Manager, right-click and choose Import Project.

2 Navigate to R18 Editors Guide/Lesson 05. Select the **OMO MULTICAM INTERVIEW.drp** project file and choose Open, and then double-click the project in the Project Manager to open it.

3 Switch to the edit page, if necessary, and then relink the media files using the Relink Media button.

This project contains footage from the interview with Chris Lang from Organ Mountain Outfitters.

4 In the timeline, review the current edit, **OMO_Multicam_EDIT**.

This simple timeline has the opening timelapse shot and logo from the Organ Mountain Outfitters promo video you edited in Lessons 1 and 2, followed by a B-roll featuring the "Buy a Shirt, Give a Lunch Forever" promise.

The director wants to insert the interview between the logo and the "Buy a Shirt…" sign.

5 Position your timeline playhead on the first frame of **SHIRT SIGN** clip.

6 In the media pool, select the Interviews bin in the OMO bin and review the footage.

This bin contains three interview clips with Chris in which he introduces himself and explains the Organ Mountain Outfitters "Buy a Shirt, Give a Lunch" concept. Thankfully, because this interview was shot on more than one camera, you can cut between the different cameras instead of adding B-roll cutaways to cover jump cuts or patching the interview with the Smooth Cut transition. However, before you start editing this interview, you need to sync the clips.

There are several ways you can sync multicamera footage on the edit page. In this first exercise, you'll manually sync the footage in a normal timeline so you can appreciate how multicam clips are organized.

7   In the media pool, right-click the Interviews bin and choose Create Timeline Using Selected Bin.

8   In the New Timeline window, name this timeline **OMO Interview**, deselect Use Selected Mark In/Out, and click Create.

A new timeline is added to the Interview bin and automatically opens in the timeline.

9   In the timeline, select the second clip, **B 0002.mov**, and drag it up to create new V2 and A2 tracks.

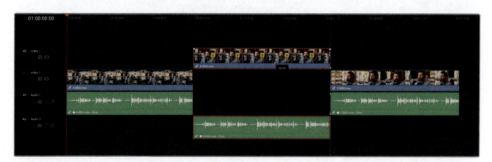

10   Move the third clip – **C 002.mov** – up to create new V3 and A3 tracks.

Next, you'll get Resolve to automatically synchronize these three clips in the timeline. There are two main ways you can do this: if you have accurate timecode across all these clips, then you can synchronize using the timecode. Alternatively, you can use the audio waveforms.

Because this footage has accurate timecode recorded across all three clips, that's what you'll use to sync the clips.

11 In the timeline, select all three clips, and then choose Clip > Auto Align Clips > Based on Timecode.

After a brief analysis, Resolve correctly synchronizes the two clips in the timeline.

Now that you have the angles synchronized, you need to convert the OMO Interview timeline into a special multicam clip.

12 In the Interviews bin, right-click the OMO Interview timeline and choose Convert Timeline to Multicam Clip.

The timeline closes (because it's no longer a timeline), and its bin icon changes to that of a multicam clip.

You're now ready to insert this interview into the main timeline and start editing between the different angles.

Viewing and Trimming the Multicam Clip

You can work with a multicam clip just like any other source clip. However, because it contains multiple angles in a single clip, you can switch between them at any time.

1   Click the Options menu (three dots) in the source viewer and choose Show Full Clip Audio Waveform.

2 Double-click the **OMO Interview** multicam clip to open it in the source viewer.

Because this is a multicam clip, Resolve automatically displays the angles simultaneously. The angle with the red outline represents the currently active angle—the angle that you will see and hear.

> **NOTE** The names of these angles are "Video 1," "Video 2," and "Video 3" because these were the names of the video tracks in the original timeline from which the multicam clip was created. You'll learn how to rename angles later in this lesson.

3 Play the multicam clip in the source viewer. All angles should be in sync.

4 In the source viewer, set an In point just before Chris says, "My name is Chris Lang..." and an Out point after Chris says, "...our communities that we live in."

5 Press F9 to Insert or click the Insert Clip button in the timeline toolbar (or however you prefer to perform an Insert edit) to insert this multicam clip into the timeline at the playhead position.

When you play this multicam clip in the timeline, you'll see only the currently active angle. The other angles are hidden.

Now that you have the interview in the timeline, you can start to edit to improve the flow.

6 If required, resize the audio track so you can easily see the waveform of the interview audio.

7 In the timeline, play through the interview until just after Chris says, "…Organ Mountain Outfitters." Press I to add and In point.

8 Continue to play through the interview until after the next question and just before Chris says, "…a lifestyle and outdoor brand." Press O to add an Out point.

9   Press Shift-Delete (or Shift-Backspace) to ripple delete the marked section.

10   Press / (slash) to preview the edit.

11   If necessary, enter Trim Edit mode and refine the edit so Chris says, "My name is Chris Lang, and I'm the founder of Organ Mountain Outfitters, a lifestyle and outdoor brand...."

12   Continue to play the interview and mark In and Out points around the breath after Chris says, "...outdoor brand" and before he says, "That not only promotes...."

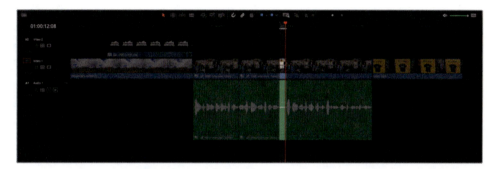

13   Again, press Shift-Delete (or Shift-Backspace) to ripple delete the marked portion of the interview, press / (slash) to review the edit, and refine using Trim Edit mode if required.

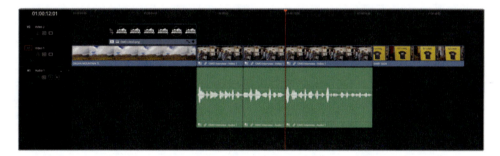

By now, this process should be second nature to you. You can treat this multicam clip exactly like any other clip you've worked with previously. However, once you have the radio edit working to your satisfaction, you're ready to put the additional angles of your multicam clip to work.

Switching Multicam Angles

Not all multicamera edits must done on-the-fly, cutting between angles in real time as the footage races past. You can switch between the different angles within a multicam clip at any time.

1   In the OMO Multicam EDIT timeline, place the playhead on the first edit in the interview.

2   Press / (slash) to review the cut.

Ouch! You'll no doubt agree that this is a rather nasty jump cut!

3   Right-click the second multicam clip in the timeline and choose Switch Multicam Clip Angle > Video 3.

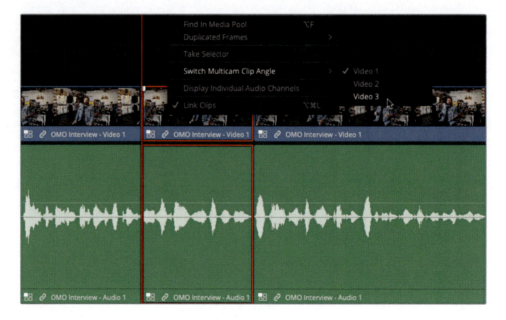

4   Press / (slash) to review the new edit.

By switching the multicam clip to the third video angle, you have successfully covered the jump cut between the two parts of the interview!

You can continue to use this technique to help tighten and make the interview more visually engaging.

5   Continue playing the multicam clip in the timeline and stop after Chris says, "... gives back...."

6   Press Command-B (macOS) or Ctrl-B (Windows) to add an edit at this point in the multicam clip.

7   Right-click the incoming clip to the right of this new edit and choose Switch Multicam Clip Angle > Video 2.

8 Using Trim Edit mode, trim the incoming multicam clip by about 21 frames, to where Chris says, "...to our communities..." to further tighten the interview.

9 Return to the start of the multicam clips in the timeline and review the changes you've just made.

> **NOTE** The above techniques will switch the video part of the multicam clip in the timeline. If you want or need to switch the audio, right-click the audio clip and choose Switch Multicam Clip Angle, where you will find the audio angles available.

Finessing and Flattening the Multicam Clips

Now that you have the interview sitting amid the rest of the footage, and utilizing the different angles, it's time to begin finessing how you may move into an out of the interview.

1 Press Shift-Command-L (macOS) or Shift-Ctrl-L (Windows) or click the Linked Selection toolbar button to disable your timeline's Linked Selection feature.

2 Make sure you are still in Trim Edit mode.

To begin with, you'll adjust the cut between the interview and the last clip in the timeline.

3 Select the outgoing video edit between the fourth multicam clip and the SHIRT SIGN clip.

4 Trim the video edit back by about 10 frames to create a slight L-cut over Chris's final words: "...live in."

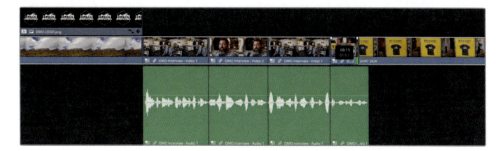

5 Press / (slash) to preview the selected edit.

That simple change has created a nice split edit that helps soften the end of the interview going back into the B-roll footage.

Now you'll turn your attention to the incoming part of the interview.

6 With Trim Edit mode still active, select the incoming side of the video of first interview clip.

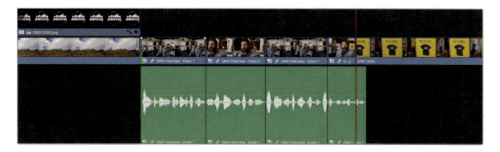

7 Position the timeline playhead over the first interview clip, at the point where Chris says his name, "...Chris Lang...."

8 Press E to make an Extend Edit and quickly create a spilt edit.

9 Press / (slash) to preview the edit.

Offsetting the edits by even a small number of frames can make a big improvement in the way the edits are perceived.

When editing multicamera footage, the final steps are to flatten the multicam clips used in the timeline to remove all the additional angles and leave just the active angles as normal timeline clips.

> **TIP** Once you've flattened a multicam clip, you will no longer be able to switch to the alternate angles. Therefore, a good technique is to duplicate the timeline prior to flattening the multicam clips so that you have an unflattened version you can change later if required.

10 Press A to switch to Selection mode and select all the multicam clips in your timeline (both audio and video). Right-click any of the selected clips and choose Flatten Multicam Clip.

Flattening the multicam clips is an important step for several reasons. First, it means that clips can be graded individually in the color page. If left unflattened, then the multicam clip itself is graded, rather than the individual angle. Second, if your multicam clips contain complex audio channel configurations, flattening the multicam clip means that you will regain access to those channels.

By shooting this interview with more than one camera, you can cut the interview effectively without having to paint it entirely with B-roll or relying on the Smooth Cut transition.

> **NOTE** If you wish to view the finished version of this edit created using the steps detailed in this section, select the TIMELINES bin, choose File > Import > Timeline and navigate to R18 Editors Guide/Lesson 05/Timelines and open OMO MULTICAM EDIT FINISHED.drt.

Editing a Multicamera Music Video

For the next multicamera exercise, you'll concentrate on editing a live music video, which will enable you to explore some of the more in-depth features of multicamera editing in the edit page. You will begin this exercise by importing a different project in the Project Manager.

1 Choose File > Project Manager or press Shift-1 to open the Project Manager window.

2 Click the Import button or right-click in an empty area of the Project Manager window and choose Import Project.

3 Navigate to R18 Editors Guide/Lesson 5, select the MISERABLE GIRL MULTICAM.drp file, and click Open.

4 Once the project has been imported, double-click it in the Project Manager to open it.

5 Click the Relink Media button and relink the offline media files.

This project contains footage of the band Jitterbug Riot performing their song "Miserable Girl."

6 Select the Miserable Girl bin and review the footage.

This bin contains eight separate video clips of the band performing their song. Each of the video clips has guide audio recorded on the in-camera microphones. A separate .wav audio file contains a clean, mixed version of the track that the band is performing.

> **TIP** You can turn audio scrubbing on and off by pressing Shift-S.

7 Click the clip **JBR_01.mov** to select it in the bin.

8 Open the Inspector and select the File tab.

The File Inspector displays some commonly used metadata for the selected clip.

Notice that this clip has Camera metadata identifying it as Camera "01" in the File Inspector.

9 At the bottom of the File Inspector, click the Next Clip button.

The next clip in the bin, JBR_02.mov, is selected and its metadata is displayed in the File Inspector.

Notice that this clip has Camera metadata identifying it as "08." The other clips in this bin are similarly identified with different Camera # metadata.

10 In the bin, select Miserable Girl Final Mix.aif.

11 Again, notice that this audio-only clip has Camera metadata identifying it as "MUSIC."

Creating the Multicam Clip

As in the previous exercise, the first step in editing multicamera footage on the edit page is to synchronize the clips. In the previous exercise, you achieved this in a manual way by arranging the clips on different tracks in a timeline and getting Resolve to synchronize them using timecode. In this example, you'll let Resolve organize the multicam clip for you using the camera metadata you just reviewed.

1 In the bin list, right-click the Miserable Girl bin and choose Create New Multicam Clip Using Selected Bin.

> **NOTE** You can also create a new multicam clip from individually selected clips instead of an entire bin.

The New Multicam Clip window opens, in which you can select how Resolve will synchronize the clips and create the multicam clip.

2 In the Multicam Clip Name field, type **MISERABLE GIRL SYNC**.

3 In the Angle Sync pop-up, choose Sound.

Selecting Sound synchronizes the clips based on their audio content.

> **NOTE** Other options for synchronizing multicam clips include using existing In or Out points and matching timecode or markers.

4 Change the Angle Name menu to Metadata – Camera.

> **NOTE** The Angle Name option dictates the order in which the angles are sorted. When you choose Sequential, Resolve labels the angles and sorts them as Angle 1, Angle 2, and so on, based on their starting timecode values, with the earliest timecode becoming the first angle. Choosing Clip Name sorts the clip names in ascending alphanumeric order and labels the angles appropriately. Similarly, choosing Metadata – Angle or Metadata – Camera sorts the clips based on information in their respective metadata fields.

5 Ensure that the option Move Source Clips to 'Original Clips' Bin is selected.

6 Click Create to create the multicam clip.

Resolve analyzes the clips' audio and, once completed, creates the new multicam clip in the selected bin. A new bin also appears called Original Clips that contains the individual camera source clips.

This multicam clip is now ready to be edited.

Real-Time Multicamera Editing

The fun aspect of working with a multicamera edit in the edit page is that you can cut between the different cameras in real time as if you were sitting in in front of a live video switcher. Often, this will save you hours because you can cut the material in the time it takes to play the timeline.

1   Double-click the **Miserable Girl SYNC** multicam clip to open it in the source viewer.

The multicam clip appears in the source viewer and displays each of the nine different clips—or angles—as a separate box, each displaying the appropriate Camera metadata information. This view organizes the angles from left to right and from top to bottom. So, the first angle (in this case, JBR_01.mov and identified as "Camera 01") appears in the upper-left window and the ninth angle (Miserable Girl Final Mix.aif and identified as "Camera MUSIC") appears in the lower-right window. Notice that the angle names have inherited the Camera # metadata information.

> **NOTE** If the playhead is at the start of the multicam clip, you will see blank frames for most of the angles because not all the cameras began rolling at the same time. Simply move the playhead to later in the multicam clip to see the footage for all the angles.

2   Play the multicam clip in the source viewer to verify that all the angles are correctly in sync.

Because Camera 01, in the upper-right quadrant of the viewer, is selected by default, that is the angle that you will hear when you begin playing the clip.

3   Return the playhead to the start of the multicam clip and set an In point just before the music starts (around 01:00:19:00).

4 Perform an Overwrite edit to edit this clip into the empty **MISERABLE GIRL EDIT** timeline, and press Home to return the playhead to the beginning of the timeline.

5 Click the Media Pool and Inspector buttons to hide the respective panels and give yourself a little more space to view the multicamera footage in all its glory.

In the timeline viewer, only the first angle, Camera 01, is displayed because that is the active camera. To be able to effectively switch camera angles and keep the source viewer in sync with the timeline clip, you need to enable multicam mode for the source viewer.

6 In the source viewer mode pop-up menu, choose Multicam to display the multicam viewer.

Enabling multicam mode replaces the transport controls under the viewer with multicam-related controls. Transport controls are not necessary in the multicam viewer because moving the timeline playhead will also move the source clips in the multicam viewer, so they maintain their synced relationship.

7   Play the timeline and see how the different angles in the multicamera viewer play in sync with the timeline viewer.

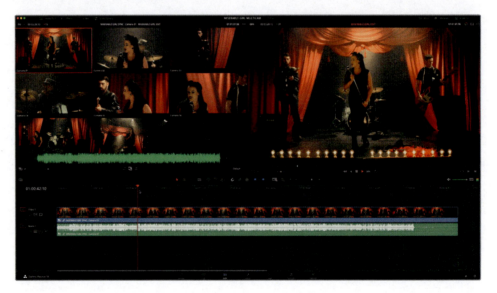

8 In the timeline, right-click the audio for the multicam clip and choose Switch Multicam Clip Angle > Camera MUSIC.

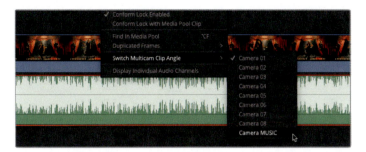

The audio for the multicam clip is changed so that it plays the audio from the Miserable Girl Final Mix.aif file as indicated by the clip's name in the timeline.

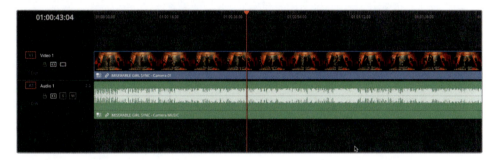

The multicam viewer also displays a green box around the MUSIC angle to indicate that this is the currently active audio.

9   In the timeline, right-click the video for the multicam clip and choose Switch Multicam Clip Angle To > Camera 04.

The video in the timeline viewer updates to display the video clip for Angle 4, and the multicam viewer now also displays a blue box around the same angle to denote that this is the currently active video angle.

You will now tell Resolve that you want to edit only the video from this multicam clip, since you don't want to edit between the different audio from the different cameras.

In the multicam viewer, three buttons control which parts of this multicam clip cut when you switch angles: video, video and sound, or sound only.

10   Click the Video button to the left to edit only the video of this multicam clip.

Now for the fun part…

11   With the playhead at the start of the timeline, start playing, and begin clicking each of the video clips in the source viewer to cut between the angles of the multicam clip in real time. As you do so, you will see the edit points appear in the timeline as you cut between the angles. Keep going until you reach the end of the timeline.

NOTE You can also press the 1, 2, and 3 keys along the top of your keyboard to cut between the angles instead of clicking on the boxes in the multicam source viewer.

12   Once you've reached the end of the song, return the playhead to the start of the timeline and click Play to review your multicamera masterpiece!

How cool is that?

Unfortunately, while it's great fun, this first attempt at a real-time multicamera edit will probably not result in a perfect edit. It's very rare that you would cut to the exact angle at exactly the right time on your first try.

Instead, think of this as your rough cut, which you will now need to refine.

Multicamera Editing Techniques

When editing multicamera material like this, don't try to make every cut perfect on the first pass. Instead, try to get a feel for the pacing of the music and the different angles you have to work with. It can be useful to watch the multicam clip back a few times before making a single cut. When you do start editing, try to make cuts according to when you feel they work best with the music, and don't be afraid to make mistakes. You'll learn how to refine the edit further in the next steps.

If you find it difficult to concentrate on so many angles at once, it's often useful to reduce the number of angles you're viewing at any one time. In this case, click the Multicam Display pop-up menu in the bottom-right of the multicam source viewer and choose an option such as "2x2." This option displays only the first four angles of the multicam clip, with the other angles then available on different "pages," which can be accessed using the Multicam page buttons that now appear.

By reducing the initial number of angles you're working with, you might find it easier to begin your multicamera edit. You can then introduce the other angles once you have the initial rough cut down. It can also help maintain real-time playback for slower systems.

You can use metadata (as you've done here) to identify the most important clips as angles 1 through 4 (or see the section "Adjusting a Multicam Clip" later in this lesson) and then order your clips appropriately when you create your own multicam clips.

Adjusting the Multicamera Edit

Now that you've completed your rough cut, when you review it, you might notice two common issues: you've made a cut at the wrong time, or you've cut to the wrong angle.

Even worse, it could be both issues at the same time!

Fear not. You're not directing live television here; this is post-production, so you can change your mind before anyone has seen your previous "creative choices." Solving the first issue where you cut at the wrong time is easy. You already know how to change an existing cut point by simply performing a rolling trim using any number of techniques.

The second issue, where you use the wrong angle, is just as easy to fix by switching angles in the source viewer.

1   In the timeline, play your multicam edit until you see a shot you want to change, and then stop playback.

    As the edit plays, the source viewer will also update because in Multicam mode the source viewer is automatically ganged to the timeline playhead position.

2   In the multicam viewer, Option-click (macOS) or Alt-click (Windows) a different image to switch the active angle to the new angle.

When you're switching the active angle, the mouse pointer changes to a replace edit icon.

It's just as easy to add further cuts to your multicam edit, too.

3 In the timeline, move the playhead to the middle of one of the multicam clips, and in the source viewer, click any other angle.

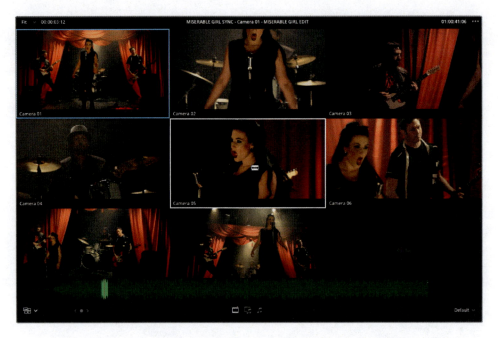

A new (through) edit point appears at the playhead position in the timeline and changes the rest of the multicam clip after the cut to the new active angle.

> **TIP** Pressing the number keys at the top of your keyboard (1, 2, 3, and so on) makes a cut at the playhead position in the timeline. Holding down Option (macOS) or Alt (Windows) and pressing a number key switches the multicam clip angle at the playhead position in the timeline. You can perform either of these operations during playback or when the playhead is stationary.

Excellent. Now you have a real taste of how much fun multicamera editing is, and how it works in Resolve. But before you start making further adjustments to this edit, you'll look at other ways you can adjust the multicam clip itself.

Multicam Editing with the Speed Editor

DaVinci Resolve 18 has improved Speed Editor functionality for multicamera editing in the edit page.

Sync Bin button activates source viewer Multicam mode

Double-press and hold to resize timeline with scroll wheel

CAM# buttons to change angles:
Press # to cut to angle number
Double-press # to switch to angle number

Adjusting a Multicam Clip

A multicamera clip is actually a type of container timeline, similar to a compound clip. You'll work more with compound clips in Lesson 7, "Compositing in the Edit Page," but for now it's useful to know that you can open a multicam clip in its own timeline when you need to make changes, such as changing the order of the angles or an angle's name.

1   In the Timeline View Options pop-up menu, select Stacked Timelines.

The current timeline appears as an individual tab along the top of the timeline window.

2   Select any of the multicam clips in the timeline and choose Clip > Open in Timeline or right-click any of the multicam clips in the timeline and choose Open in Timeline.

The multicam clip opens in its own tabbed timeline.

3   To view such a large number of timeline tracks effectively, reduce the video and audio track heights to their minimum and increase the size of the timeline window by dragging the line between the viewers and the toolbar area upward.

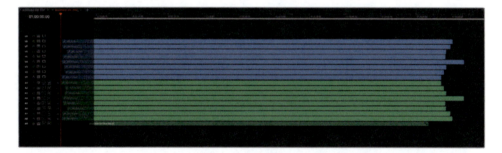

You can see how Resolve has structured and organized the multicam clip; it's very similar to the way you arranged the clips of Chris's interview in the timeline before creating the first multicam clip in this lesson. The organization is simple: any content on the video and audio 1 tracks is displayed as the first angle in the multicam viewer. If content is present on the video and/or audio 2 tracks, it is displayed as angle 2, and so on. Notice that the track names follow the names of the angles as displayed in the Multicam source viewer.

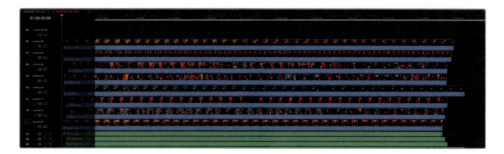

You will now add an additional angle and change the order of the angles in this multicam clip.

4   Right-click any of the timeline video track headers and choose Add Angle.

A new angle is added, V10 and A10, both called "Angle 10."

5   Open the media pool and select the Extra Angle bin.

A tenth, currently unused angle, is in this bin, called JBR_09.mov.

6 Drag the **JBR_09.mov** clip into the new V10 in the multicam clip timeline.

7 Close the media pool.

You will now need to sync this new angle to the other angles in this multicam clip timeline.

8 Ensure that Link Selection is enabled, select the clip on V10/A10, and Shift-click any other clip in the timeline.

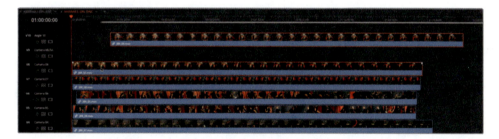

9 Right-click on any of the video clips and choose Auto Align Clips > Based on Waveform.

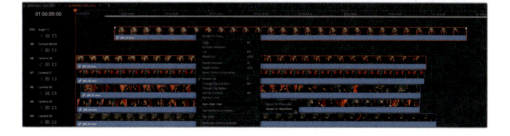

Resolve instantly syncs the new angle to the existing angles!

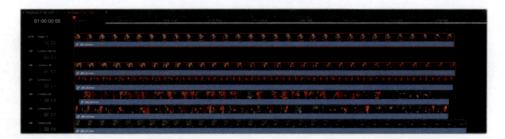

You will now rename the new angle to ensure that you keep your multicam clip organized and tidy.

10 Right-click the track header for V10 and choose Move Angle Down.

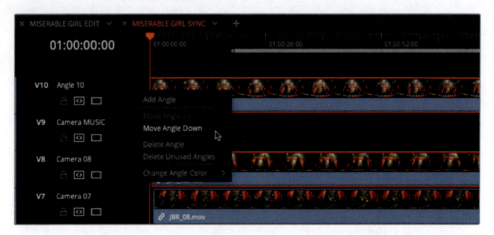

Resolves changes the order of angles V10 and V9 (as well as A10 and A9).

However, notice that this new angle is still named Angle 10.

11 Click the angle name and type **Camera 09** in keeping with the other angle names for this multicam clip.

Now you can start to use this new angle in your multicamera edit.

12 Middle-click the timeline tab for the MISERABLE GIRL SYNC to close the multicam clip's timeline and return to the main timeline for this edit.

13 Resize the timeline window to make it shorter now that you no longer have lots of tracks to view.

14 In the source viewer pop-up menu, choose Multicam to switch the source viewer to the multicam viewer.

The multicam viewer now displays the additional camera angle as Camera 09.

15 In the Multicam view pop-up, choose 3x3 to view all nine video angles in the multicam clip. The additional "MUSIC" angle can be accessed using the multicam page controls that appear to the left of the multicam view pop-up when the total number of angles is more than the current view allows.

> **NOTE** Even though your multicam clips can contain an unlimited number of angles, the maximum number of angles you can view at the same time is 25 using the 5x5 view. Any more angles in your multicam clip will be shown in subsequent "pages" in the multicam viewer.

16 Finally, watch your multicam edit through again, making any further adjustments you think are necessary. Don't forget to include the new camera angle and flatten the multicam clips once you've finished!

> **NOTE** If you wish to view finished versions of this edit created using the steps detailed in this lesson, select the TIMELINES bin, choose File > Import > Timeline, and navigate to R18 Editors Guide/Lesson 05/Timelines. Here you will find two versions of the finished timeline. MISERABLE GIRL FINISHED.drt is an edited version that you can use to practice your multicamera editing skills. This timeline will also import the associated multicam clip used in this edit because it's a different multicam clip than the one you've created as part of this lesson. MISERABLE GIRL FINISHED_FLATTENED.drt is the same edit but has been flattened so the alternative angles are no longer available.

Congratulations! You should now have the skills necessary to tackle even the most complex multicamera editing tasks in DaVinci Resolve's edit page.

Remember, when editing multicamera projects and playing back in real time, you are really trying to assess the rhythms and themes of the music and capture those characteristics as you are cutting. Sometimes you might cut a multicam clip in three or four ways to experiment with various pacing strategies and later decide which one is showing the most love. However you approach your multicamera projects, though, as with all editing, each cut requires constant revisiting and reworking to ensure that your audience sees the best possible results.

Lesson Review

1   What options do you have for synchronizing angles in a multicam clip from video clips without sound?

     a)   In or Out points

     b)   Markers

     c)   Timecode

2   What is the maximum number of angles you can view simultaneously in a multicam clip?

     a)   9

     b)   16

     c)   25

3   Which modifier key is held down to switch the entire multicam clip to another angle instead of adding a new edit point?

     a)   Command (macOS) or Ctrl (Windows)

     b)   Option (macOS) or Alt (Windows)

     c)   Shift

4   True or false? You cannot change the angle names, add additional angles to, or change the current order of the angles of an existing multicam clip.

5   True or false? When you choose to flatten a multicam clip, you lose all the other synchronized angles from the multicam clip(s) in the timeline.

Answers

1   a), b), and c) You can choose to synchronize angles using In points, Out points, timecode, or markers instead of sound.

2   c) 25. You can have multiple pages of more angles, but the maximum number of angles you can view in any one page is 25 (5x5).

3   b) Option-click (macOS) or Alt-click (Windows) an angle to switch the existing angle.

4   False. Right-click a multicam clip and choose Open in Timeline to adjust an existing multicam clip.

5   True. Flattening a multicam clip removes all the unused angles and leaves only the clip that was used as the active angle in the timeline.

Project Organization

While DaVinci Resolve is a superior editing, audio mixing, visual effects, and color grading system, it can also play a key role on set before a single edit is made. In this lesson, you'll focus on some of the powerful yet lesser-known and often overlooked functions that will help you organize and optimize high-resolution, camera-original media and generally prepare everything for your edit.

Time

This lesson takes approximately 60 minutes to complete.

Goals

With that in mind, DaVinci Resolve has a specific page dedicated to helping you focus on preparing and organizing the media for your project: the media page. It is here that you have the most space to explore the clips you have to work with, without the distraction of timeline windows, color grading controls or audio mixers.

Yet project organization isn't something that happens only after you import the clips into your project; rather, it's something that's constantly being refined as you work on a project, often right up until the point you're ready to deliver it. As such, although this lesson is designed to highlight this functionality in the media page, many of the steps discussed in this lesson can be applied directly in the edit page so you can implement them alongside your editing.

In order to explore these steps so that you can begin applying them to your own projects, you'll set up a new project and import the Organ Mountain Outfitters' footage you worked on in Lessons 1 and 2.

Creating a New Project and Project Settings

The first step for any project is to actually create a new project in the Project Manager.

1   Open DaVinci Resolve and, in the Project Manager window, click the New Project button.

2   In the Create New Project window, type **My Project** as the name for this new project.

3   Click Create.

4   Click the Media page button and choose Workspace > Reset UI Layout.

As mentioned previously in the introduction to this chapter, the media page is dedicated to the reviewing and organizing of your source media—principally video, audio, and graphics files.

Media Storage Locations   Media Storage Browser   Viewer   Audio

Bin Lists   Media Pool   Metadata

When you create a new project in DaVinci Resolve, the project itself has some default settings it draws upon. The next step when creating a new project should be to check these basic settings.

5   Choose File > Project Settings or press Shift-9 to open the Project Settings window.

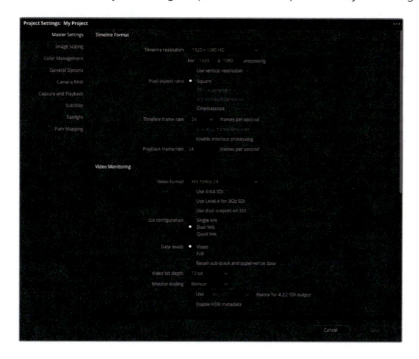

As an editor, the main settings to be concerned with are in the Master Settings. This group of settings primarily deals with the Timeline Settings and Video Monitoring for this project.

By default, all projects have a timeline resolution of 1920 x 1080 HD and a timeline frame rate of 24 frames per second (fps). This means that this is the resolution and frame rate used whenever you create a new timeline using the Project Settings option. These settings can be changed when you create a new timeline, so each timeline in any project can have completely different settings, but as most projects tend to use the same timeline settings, it's useful to set a general setting in the Project Settings window.

So, what timeline settings should you use? Well, that depends primarily on where you will deliver your final edited timeline. For example, if you're working on a feature-film project, you'll probably want to deliver a 3840 x 2160 Ultra HD timeline at 24 frames per second; if it's a broadcast TV show, then the timeline might be 1920 x 1080 HD at 29.97 frames per second (or 25 frames per second for the UK and Europe); whereas if you will deliver a file to a streaming site such as YouTube or Vimeo, then your choices aren't quite so limited and you may need to make a judgement call based on the source footage itself. If in doubt, it's always worth talking to the director to determine the best settings for your project.

> **NOTE** If you're working for broadcast TV, you may still be required to work with an interlaced timeline, which DaVinci Resolve fully supports. To enable interlaced processing for a timeline, select the "Enable interlace processing" option in the Timeline Format section of Project Settings or in the Format tab of the New Timeline window. The frame rates for interlaced timelines are measured in fields per second and can only be set to 50, 59.94, or 60, equating to 25, 29.97 and 30 frames per second, respectively. Everything in an interlaced timeline, including graphics, Fusion compositions, and video clips, are processed at the field level for high-quality compositing and titling for interlace delivery.

Although the footage for this project for Organ Mountain Outfitters, which is destined to be uploaded to different streaming sites, was shot at different resolutions and frame rates, the main footage was shot at 1920 x 1080 HD and at 23.976 frames per second. Therefore, this is the resolution and frame rate you will work with for this project.

6   Select 1920 x 1080 HD from the Timeline Resolution dropdown menu and 23.976 frames per second from the Timeline Frame Rate dropdown menu.

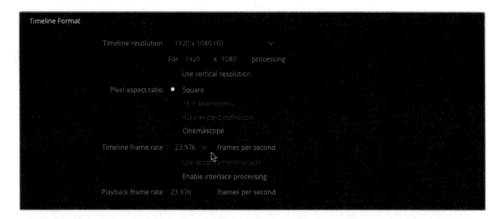

> **NOTE** When you change the frames per second option, the Playback frame rate and Video Format in the Video Monitoring section change to match. Normally, this is what you would want; however, in very rare instances you can adjust these separately, especially if you're working with a timeline resolution or frame rate that is unsupported by the video monitoring hardware.

Saving Project Presets and the Setting Default Preset

If you frequently work with projects that require different Timeline Format settings or you need a specific setting for your timelines for all projects beyond the standard 1920 x 1080 HD and 24 fps, it can be useful to save different project presets that can be easily recalled, or even set the current project preset as a default for all newly created projects in the current Project Library.

1   In the Project Settings window, click the Options (three dots) menu and choose Save Current Settings as Preset.

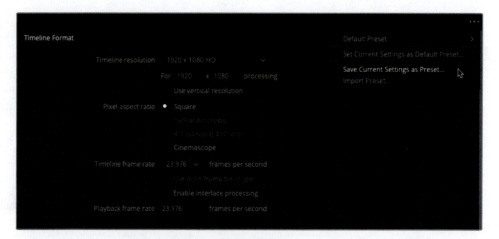

2   In the Preset Name window, type **1080p23.976** and click OK.

All the settings for the current project are now saved as a preset you can quickly load.

3   Click the options menu again and choose Default Preset > Load Preset to reload the default project settings, returning the Timeline Format settings to their starting values.

4   Click the options menu and choose 1080p23.976 > Load Preset to reload the saved preset you'll use for this project.

> **NOTE** Saving a project preset saves all the settings in the Project Settings window, not just those in the Master Settings (a full discussion of which is beyond the scope of this editing training guide).

You can also export a project preset that you can use to quickly load the same project settings onto another DaVinci Resolve 18 system.

5 Click the options menu and choose 1080p23.976 > Export Preset.

6 In the Save As field, type **1080p23.976**, choose a location, and click Save to save the exported .preset file.

> **NOTE** To import this project .preset file onto another DaVinci Resolve 18 system, select the options menu in Project Settings and choose Import Preset.

You can also set any saved preset as a default for all new DaVinci Resolve projects for the current Project Library.

7 Click the Project Settings options menu and choose 1080p23.976 > Set as Default Preset.

A window appears confirming that you want to set the 1080p23.976 preset as the default for all future projects.

8 Click Set.

Now all future projects created in the current Project Library will use this preset by default.

Importing Media

There are numerous ways to import files into a DaVinci Resolve project. For instance, you could choose File > Import > Media on any page that has access to the media pool (that is, every page except the deliver page). You can also simply drag and drop files from the Finder (macOS) or File Explorer (Windows) directly into the media pool! However, both of these techniques offer limited options, whereas the media page has much more flexibility when it comes to importing clips.

> **NOTE** Whenever you import media, DaVinci Resolve creates a link to the original clips on your hard drive. At no point does this import process copy, move, convert, or in any way alter the source media.

1   In the media storage, navigate to the R18 Editors Guide > Media > Organ Mountain Outfitters folder.

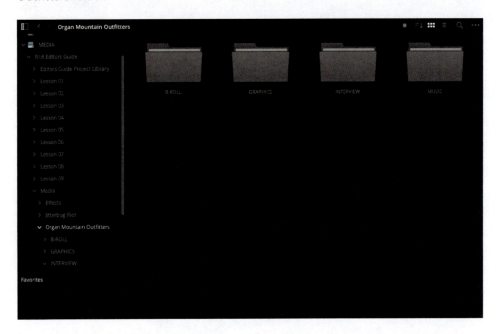

The media storage browser displays four folders containing the clips used in the Organ Mountain Outfitter's promo.

2   Double-click to open the INTERVIEW folder.

This folder contains two additional folders, AUDIO and VIDEO.

3 Open the VIDEO folder and move your mouse across any of the clips to Live Preview in the viewer.

> **TIP** You can disable and enable Live Preview in the viewer's options (three dots) menu.

4 Select any of the video clips to display the clip in the viewer.

When a clip is selected, the Metadata panel gives you some information regarding that clip, including the file's codec, resolution, frame rate, and number of audio channels.

5 Press Spacebar or L to play the clip in the viewer.

As the clip plays back, the Audio panel displays the levels of the embedded audio channels.

> **NOTE** The media page supports the same keyboard shortcuts for playing clips that you've been using on the edit page.

6 In the media storage browser, click the Back button twice to return to the Organ Mountain Outfitters folder level.

7 Double-click the B-ROLL folder to display the clips inside that folder.

8 From the top of the browser, click the List View button.

The List View displays more information about the individual files than the Thumbnail View does. Here, you can see at a glance that these clips have a resolution of 3840 x 2160 and frame rates of either 23.976 or 59.94 fps. Therefore, using the media storage browser, you can assess the footage you have prior to import and even choose which specific clips you want to import into your project. Other import methods aren't quite so helpful.

9   Click the Back button to return to the previous folder level.

10 Select the four folders in the Organ Mountain Outfitter's folder.

11 Right-click any of the selected folders to reveal the importing options.

There are three main options for importing clips from the media storage browser:

— **Add Folder into Media Pool** will import the contents from one level inside the current folder.

— **Add Folder and SubFolders into Media Pool** will import the contents of the current folder and any folders contained within it. This is the same as when you drag a folder into the media pool either from the media storage browser, or from the Finder (macOS) or File Explorer (Windows). This is a useful option when importing numerous files from different folders from a camera card that uses a complex directory structure.

— **Add Folder and SubFolders into Media Pool (Create Bins)** will import the contents of the current folder and any folders contained within it, preserving the folder structure as a series of bins in the media pool. This option is most useful when you want to import several clips that are already organized into folders on your hard drive.

12 Choose Add Folder and SubFolders into Media Pool (Create Bins).

> **NOTE** If Resolve presents a dialog asking whether you want to change the project frame rate and you have already set the current project to 23.976 frames per second, click Don't Change; otherwise, Resolve will change the project frame rate to 59.94 frames per second!

The clips are added to the media pool in a series of bins that reflect the folder structure of the Organ Mountain Outfitters folder.

Generating Proxy Files

Modern digital video files are highly complex assets that contain huge amounts of information, as attested to by their file sizes, with many individual files easily measuring in the tens of gigabytes (GBs) and many projects in the terabytes (TBs) in size!

Working with camera-original content is ideal when color grading but, when it comes to editing, these large, complex files can slow you down if they overtax the hardware you're working on. As you try out different shots, trimming and adjusting clips, you need a proper feel for the pacing of a scene and the changes you're making. A computer that cannot process media efficiently, or a drive that is not fast enough to play high-resolution or high-frame rate media, can result in a frustrating editing experience.

Resolve includes a convenient method for creating lower-resolution clips as proxy media while retaining a relationship with the camera originals. Generating proxy media enables the speed you want when editing yet leaves you only one click away from the camera-original media when you need it for color grading or VFX work.

You can choose to generate proxy files directly from DaVinci Resolve or by using the Blackmagic Proxy Generator. Both have advantages, depending on your workflow.

Using the Blackmagic Proxy Generator

The Blackmagic Proxy Generator is a separate program that is installed alongside DaVinci Resolve and can automatically create proxy media from source video files placed in a designated watch folder. The Proxy Generator is a lightweight application that can be left running in the background, and DaVinci Resolve will automatically recognize and use the proxy media as it becomes available. Moreover, the Proxy Generator can be used to manage proxy files quickly and effortlessly.

> **NOTE** If you're using the free version of DaVinci Resolve, the Proxy Generator program will be called "Blackmagic Proxy Generator Lite."

1   Open the Blackmagic Proxy Generator application. If a file browser window also opens, click Cancel.

The Proxy Generator is a simple program that has no complex settings.

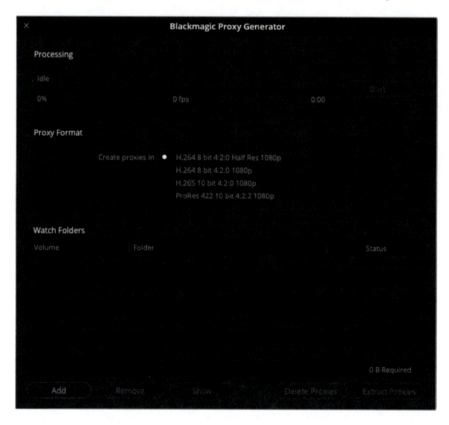

The first step in using the Proxy Generator is to set a "watch folder." A watch folder is simply a normal folder on your system that contains your source media files.

2   Click the Add button to add a watch folder location.

3   Navigate to R18 Editors Guide/Media/Organ Mountain Outfitters/B-ROLL and click Open.

The watch folder is added to the list, detailing the volume, the specific folder, and the current status of the folder. You can add as many watch folders to this list as required.

> **TIP** You can drag folders directly into the list of watch folders from the Finder (macOS) or File Explorer (Windows).

You can choose one of four preset settings (three for Windows users) for your proxy media. A value at the bottom right of the list of watch folders indicates the amount of storage required to generate the proxies for the selected location(s).

4   In the Proxy Format section, choose the H.264 8 bit 4:2:0 Half Res 1080p option.

This option will create proxy media at one-half the resolution of 1080 HD media (960 x 540) and using the H.264 video codec. This will be useful for editing but not for grading, mainly due to the 8-bit and 4:2:0 chroma subsampling properties of these files.

5 Click Start.

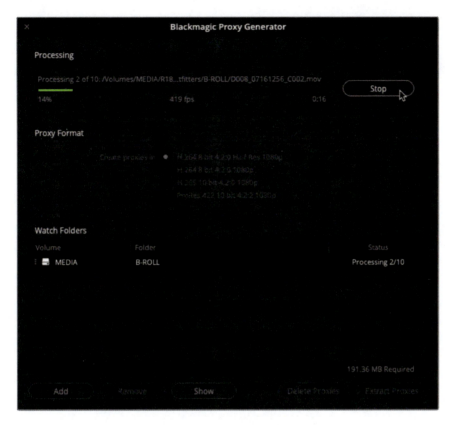

The Proxy Generator begins processing the source video clips in the watch folder, placing the generated proxy media in a Proxy subfolder in the same location as the source media. This allows Resolve to automatically see the relevant proxy media for clips in this source folder.

6 Return to DaVinci Resolve and select the B-ROLL bin in the media pool.

The clips in this bin have a new icon indicating that these clips have proxy media associated with them. This icon will remain displayed on any clip in the media pool, the edit page, or color page timelines whenever Resolve is using the proxy media of a clip, as opposed to the original source file.

7   In the media pool, click the List View.

8   Right-click the top of the Clip Name column and, in the list that appears, select the Proxy column.

The Proxy column appears to the right of the Clip Name column and details the proxy file resolution. Resolve will now use the lower resolution H.264 proxy file when playing these clips, which means your real-time performance will increase, allowing you to focus on the editing at hand.

Generating proxy media using the Proxy Generator is efficient and easy, without any complicated set up needed. Using lower-quality proxy media files will help improve your editing workflow, especially when using very high-resolution or high frame rate source media. However, you don't want to inadvertently use the proxy media when grading, but you can easily switch from the proxy media back to the original files in the Playback menu.

9   Choose Playback > Proxy Handling > Disable All Proxies.

The icons for the clips in the media pool return to normal, indicating that the proxy files are no longer being used, but the proxy media remains available.

10 Choose Playback > Proxy Handling > Prefer Proxies to re-enable the proxies for the project.

You can also use the Proxy Generator to manage your proxy media.

11 Return to the Blackmagic Proxy Generator application and click Stop.

12 Select the B-ROLL folder in the list of watch folders and click Show.

A Finder window (macOS) or a File Explorer window (Windows) opens to reveal the B-ROLL folder.

13 Open the folder to reveal the Proxy subfolder with the generated proxy media files.

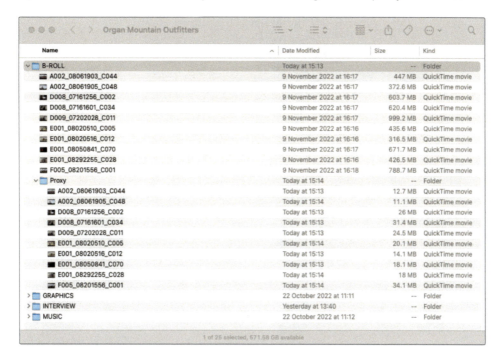

14 In the Proxy Generator, click Delete Proxies.

15 A warning appears confirming that you want to delete all the proxy folders and clips.

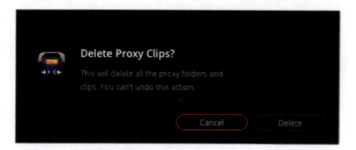

Don't worry! This warning refers to the proxy media files and folders only. Your original source clips are safe from being deleted.

16 Click Delete, and then click Done.

The proxy media and subfolder are instantly removed from your system.

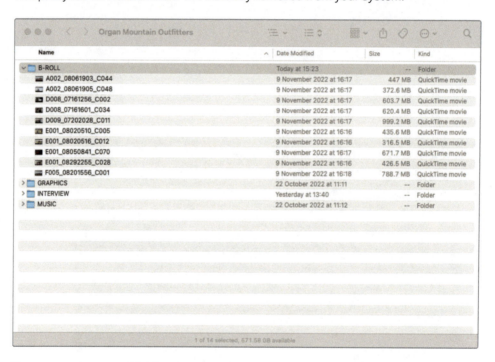

17 To recreate the proxy files for the watch folder, click Start.

> **NOTE** The Proxy Generator also has a button to Extract Proxies. This will copy the proxies for the selected watch folder(s) to a location of your choice. This is useful for creating a separate proxy-only folder that you can hand over to another editor via a portable hard drive or cloud storage.

Generating Proxy Media from the Media Pool

The Proxy Generator is a handy, intuitive way to quickly create proxy media. However, an alternative way of creating proxy media is directly from the media pool in Resolve. This can be useful if you only need to generate proxy media for a limited number of clips since the Proxy Generator application will generate proxies for all the media in a watch folder.

1   In Davinci Resolve, press Shift-9 to open the Project Settings.

2   In the Master Settings, scroll down to the Optimized Media and Render Cache group of settings, which also includes settings for creating proxy media.

3   Change the Proxy media resolution dropdown menu to Quarter.

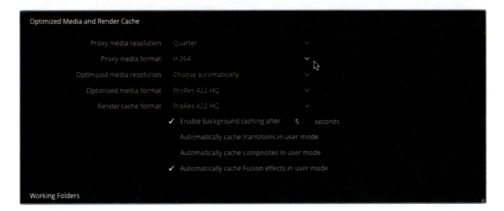

This will create proxy media that is one-quarter the resolution of the source media file. This setting is relative, so proxies for 1080 HD media will be 480 x 720, and proxies for UHD clips will be 960 x 540.

4　Change the "Proxy media format" dropdown menu to H.264.

5　Scroll down to the Working Folders section.

The "Proxy generation location" specifies where the proxy media will be created for this project. You will leave this unchanged.

6　Click Save to save the changes to the Project Settings.

7　Choose DaVinci Resolve > Preferences.

The System Preferences window opens to the media storage settings. Here, you'll find three "Proxy Generation Location" options:

— **Proxy subfolders in media file locations** will create the proxy media in a "Proxy" folder in the same location as the source media file.

— **Use project setting** creates the proxy media in the location specified in the Project Settings.

— **Ask when creating** allows you to manually select the location where the proxy media will be created.

> **NOTE** Choosing the "Proxy subfolders in media file locations" generates the proxy files in the same location as the Proxy Generator application does. Choosing this option also means that the proxy files can be deleted and extracted by the Proxy Generator if the enclosing folder is added as a watch folder.

8   Choose "Proxy subfolders in media file locations" and click Save.

9   Select all the clips in the B-ROLL bin, right-click and choose Generate Proxy Media.

The proxy media will be generated for the clips. Unlike using the Proxy Generator, though, you will have to wait until all the proxies have been generated before you can continue working.

Syncing Audio to Video

Now that you've imported the rushes into your project and generated proxy media, you can really start to organize the footage by syncing any audio and video clips that were recorded on separate devices.

Many productions record audio on dedicated digital audio devices to capture the highest quality audio or when it's not practical or desirable to record audio directly to a camera. Thus, when the files come in from the day's shoot, you'll need to sync the appropriate audio and video clips—a process often referred to as "syncing the dailies." Thankfully, Resolve has a fantastic way of making this process as painless as possible. And, similar to the multicamera syncing you did in the previous lesson, you can use either the timecode or the audio of the clips to help achieve the perfect sync!

1   In the bin list, select the AUDIO and VIDEO bins inside the INTERVIEW bin to display the contents of both bins in the media pool.

2   Play the first clip, A 0002.mov, in the viewer.

This is one of the interview clips with Chris Lang that you worked with in Lesson 1. Unfortunately, the audio isn't very well recorded. From the meters in the Audio panel, you can see that this clip has two audio channels that are very low.

3   Right-click the clip and choose Clip Attributes.

The Clip Attributes window for the selected clip opens. This is where you can alter certain video, audio, timecode, and naming properties of a clip or clips.

4   Select the Audio tab.

As you can see, the audio tab of clip attributes tells you that this clip has two embedded audio channels, configured as a stereo audio track.

5 Click Cancel to close the Clip Attributes window without saving any changes.

6 Select the first audio clip, **A-002.WAV**, and play it.

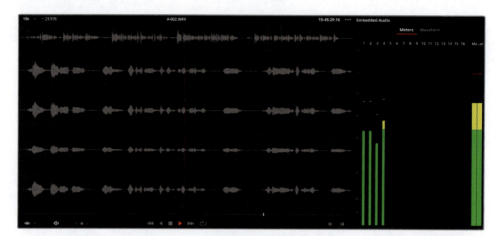

This audio clip has four audio channels, displayed as separate waveforms in the audio viewer, all with healthy audio levels showing in the meters.

7 Right-click the audio clip and choose Clip Attributes.

The Clip Attributes window opens, displaying the Audio tab. This clip has four individual mono channels labelled "Mix-L," "Mix-R," "Boom," and "Chris." This indicates that a boom mic was recorded on channel 3, Chris's personal mic was recorded on channel 4, and channels 1 and 2 are a mix of both.

> **NOTE** The track names have been imported as part of the audio files' metadata. You can view specific track names (or add your own) by choosing Audio Tracks from the Metadata panel's Sort menu.

8   Click Cancel to close the Clip Attributes without making any changes.

Now that you understand how the audio for these clips has been recorded, you can sync them together and choose the most appropriate channels to work with.

9   In the media pool, select all the clips in the selected bins, right-click them, and choose Auto Sync Audio > Based on Waveform and Append Tracks.

After a short analysis, Resolve has synchronized the audio with the video clips!

10   Select the first video clip again, A 0002.mov, and play it.

Now this clip has six audio channels, the last four of which are much higher in the meters than the first two.

> **TIP** You can verify that there is synced audio with each of the clips in List View in the Synced Audio column.

Modifying Audio Channels in Clip Attributes

You could continue using the audio of these clips as is. Having multiple audio tracks on a clip gives you the opportunity to choose which microphone to use at any given time. However, it can be just as simple to configure the audio of a clip so that it is using just the track(s) you need.

1   In the bin list, Command-click (macOS) or Ctrl-click (Windows) the AUDIO bin to deselect it, leaving just the synced interview clips in the media pool.

2   Select all the interview clips, right-click them, and choose Clip Attributes.

Now the Audio tab of the Clip Attributes window is showing that these clips all have a total of six audio channels: the first two being the embedded audio channels, and the last four being the linked audio clips. Since you only need the audio recorded from Chris's personal mic, you can remove the audio channels you don't need.

3   Click the trash can icon of the embedded stereo track to remove it.

4   Repeat for the next three linked mono tracks—Mix-L, Mix-R, and Boom.

This leaves just Chris's personal mic as the only remaining mono audio track.

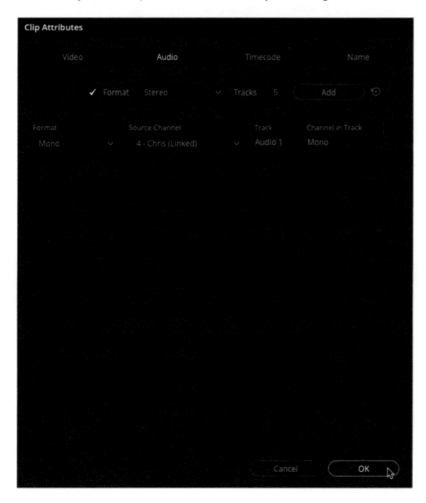

5 Click OK to accept the changes and close the Clip Attributes window.

The interview clips are now correctly configured with the correct mono audio channel.

> **NOTE** The process of removing the channels from the Clip Attributes has not "deleted" the audio at all. Instead, it simply disables the tracks for the clip, allowing you to reconfigure them again at any time by reopening the clip's Clip Attributes.

While Clip Attributes encompasses several useful configuration features, ideally, you'll want to configure most of them before you edit a clip into a timeline. Once clips are placed into a timeline, any changes you make to a clip's attributes in the media pool will only affect new instances of that clip added to a timeline, with existing instances of the clip in any timeline remaining unchanged. However, you can adjust these existing clips within the timeline by right-clicking the clip and choosing Clip Attributes from the contextual menu.

Configuring Metadata Presets

Metadata has quickly become an important part of the editing process, not least in the sorting and finding of specific clips in the morass of media in any given project. However, as useful as metadata is, it can often seem overwhelming at first.

Resolve has several metadata categories you can use to reduce the amount of metadata displayed at any one time to a manageable subset of the whole. However, you can customize metadata presets to display only the information you most need or want to see.

1 Choose DaVinci Resolve > Preferences, or press Command-, (comma) in macOS or Ctrl-, (comma) in Windows.

2   In the Preferences window, click the User tab and select the Metadata category to the left.

Using the Metadata Presets, you can create, modify, and delete custom metadata presets.

3   Click the New button to create a new metadata preset, and name it **Favorite Metadata**. Click OK.

The newly created Favorite Metadata preset appears in the Metadata Presets list.

In the lower half of the Metadata pane, under Metadata Options, you'll see all the metadata you can add to the preset.

4   Select the checkboxes for Description, Keywords, Comments, Camera #, People, Scene, and Shot. These are the metadata fields you will need throughout the rest of this lesson.

5   At the upper right of the Metadata options, click Save to save your changes to
     the preset.

6   Click Save at the bottom of the Preferences window to save and close the
     Preferences panel.

7   In the media pool, select any clip from the B-ROLL bin.

8   In the Metadata panel, click the Sort menu and choose All Groups.

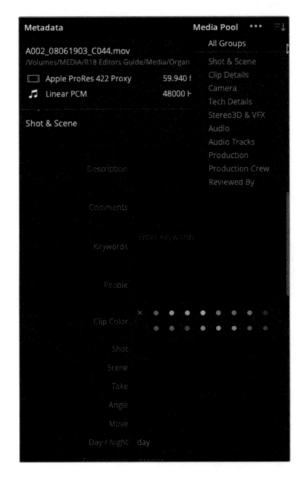

9 Next, click the options menu (three dots) and choose the new Favorite Metadata preset.

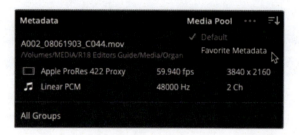

> **TIP** Your preset will appear blank in the Metadata Editor unless it is set to show all groups or a subset that contains the fields of your preset.

Only the selected metadata fields for the Favorite Metadata preset appear in the Metadata Editor.

You could continue entering the metadata manually or, if the information exists outside of DaVinci Resolve, you could simply import it.

Importing and Using Metadata

You have many ways to populate your clips with useful metadata. It may be entered on the camera during production (although detailed metadata is rarely a priority for the camera operator or an assistant); you can enter it manually yourself, which very few people want (or have the time) to do; or someone on set can be assigned to be responsible for entering metadata in their favorite spreadsheet program or in any of the smart slate apps that can be used to log metadata such as shot, scene, take, and more. You can then import this data into Resolve using the simple CSV (comma separated value) format and save yourself hours of work in the cutting room!

1   Choose File > Import Metadata To > Media Pool.

2   In the File dialog, navigate to R18 Editors Guide/Lesson 06 and select the file **OMO metadata.csv**.

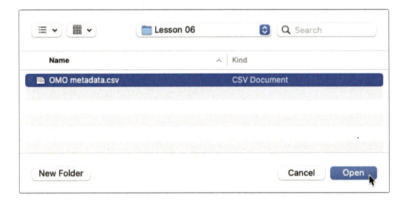

This .csv file was exported from a simple spreadsheet program.

3   Click Open.

The Metadata Import dialog opens. This window allows you to choose how you want Resolve to match the clips with additional metadata. In this case, you can match clips based on their filenames but not their timecodes because that information is not included in the .csv file you're importing.

4   Deselect the "Match using clip start and end Timecode" checkbox since the .csv file does not contain this information.

5   In the Merge Options, select "Update all metadata fields available in the source file."

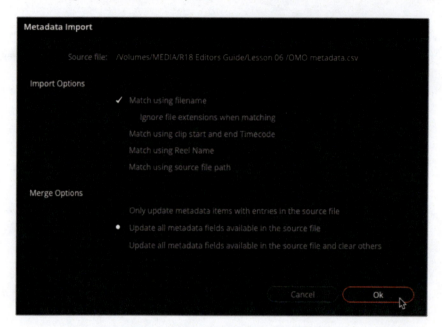

Importing metadata in this manner will replace any existing metadata in the fields detailed in the .csv file. Metadata fields not listed in the .csv file will remain unchanged.

6   Click OK.

A confirmation window appears stating that the information within the .csv file has been imported and added to 24 clips.

7   Verify that the information from the .csv file has been added to the media pool clips by selecting a few clips and viewing their newly added metadata in the Metadata panel.

Your clips now include scene, shot, description, and keyword information. This metadata will help as you organize and rename the clips to something more useful than the cryptic filenames given to them by the camera.

Exporting Metadata and Bins

In addition to importing metadata to clips in your media pool, as detailed in the preceding steps, you can also choose to export the metadata from your media pool or from selected clips from the media pool. To do so, simply select File > Export Metadata From > Media Pool, or File > Export Metadata From > Selected Media Pool Clips (as appropriate). Your chosen clip metadata will then be exported as a .csv file and will provide a way of easily transferring metadata from one project to another that uses the same media, even if that project is on another Resolve system.

You can also export an entire bin by choosing File > Export > Export Bin. This command will export the clip metadata (not the media) from the currently selected bin to a .drb file. As with exported metadata, you can use this option to transfer bins between different Resolve projects or systems by choosing File > Import > Import Bin.

Any clips listed in the .drb file will be automatically imported into the current project, together with their associated metadata. If the source media is in the same location as the system the bin was exported from, it will be automatically linked to the newly imported clip. If the media is in a different location, you will have to manually relink it.

Searching Using Metadata

You also can use metadata to find clips quickly and easily. Being able to find the material you want or need as rapidly as possible means that you can more effectively focus on the story and the flow of your edit.

1   In the bin list, select the Master bin.

2   At the top of the media pool, click the Search button (the magnifying glass) to reveal the search field.

By default, the search will only look in the current bin and will search only across clips' filenames.

3   Click the Search menu and choose All Bins.

4   In the "Filter by" dropdown menu to the right of the Search field, choose All Fields.

Resolve will now search all the bins and across all the available metadata fields.

5   In the search bar, type **chris** to display all the shots that have "Chris" as part of their metadata.

6 In the search bar, highlight "chris" and type **Pine** to reveal two shots that include "pine" somewhere in their metadata (in this case, it's from the "PINE TRAIL" in the scene field).

7 Click the Search button again to close and clear the search field.

Resolve's powerful and responsive search feature lets you leverage the flexibility of metadata to instantly find media pool clips in even the largest project.

Using Automatic Smart Bins

Another advantage of adding metadata to your clips is that you can use it to create Smart Bins. Keywords, scene, and shot metadata can be used to automatically create a series of Smart Bins.

1 In the Smart Bins area of the media pool, click the disclosure arrow for the Keywords Smart Bin folder to view the list of Smart Bins.

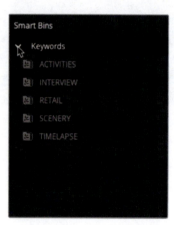

These Smart Bins were automatically created based on the keywords applied to each clip in this project. If you adjust the Keyword metadata of any clips within this project, this list of Smart Bins will update accordingly.

To show additional automatic smart bins for other metadata, you need to open the User Preferences.

2 Choose DaVinci Resolve > Preferences, click the User tab, and then click the Editing category on the left side of the window.

The editing preferences allow you to display other automatic Smart Bins beyond Keywords, such as scene, shot, and people metadata.

3 Select the Automatic Smart Bin for Scene metadata and People metadata checkboxes, and then click Save to save the change and close the Preferences window.

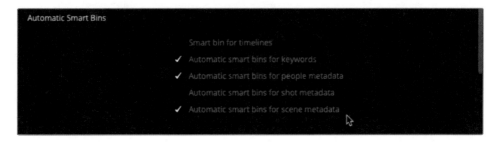

Scene and People folders appears in the Smart Bins list.

4 Click the disclosure arrow for the Scene smart bins to reveal the appropriate scene metadata from the clips in the project.

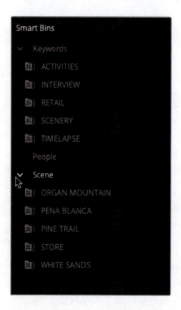

> **TIP** You can drag and drop clips onto an existing automatic smart bin to quickly add the clips to that bin by automatically assigning the metadata properties of the smart bin to the clips. For example, dragging a clip onto a keyword smart bin will automatically add the keyword to the clip.

Analyzing Clips for People (Studio Only)

> **NOTE** Smart Bins for People Metadata is available only in DaVinci Resolve Studio. If you're using the free version of DaVinci Resolve, you may read over this exercise, but you won't be able to perform the steps.

Another subset of metadata that you may find useful when organizing clips in your project is to have Resolve analyze clips for people's faces. This can then be used to generate People Keywords, which in turn will appear as a series of automatic smart bins and can be useful in locating clips containing certain people, whether this is an actor, interviewee, or presenter.

1 Select all the clips in the B-ROLL bin, right-click, and choose Analyze Clips for People.

The clips are analyzed by the DaVinci Resolve Neural Engine, the results of which are stored in the project's Face database.

Once the analysis is complete, the dominant faces that have been identified are displayed in the People window, allowing you to manage the People metadata.

2   In the People window, select the clips identified as "Person 1."

This collection contains three clips featuring the same person, with a red box identifying the specific face that the analysis has identified. You can add a name for this person to make it easier to identify them.

3   Return to the People collection and select the first icon displaying the face of "Person 1."

4   Click the area underneath the selected face where it says "Enter name here" and type **Bobby** as the guy's name.

You can also remove a face that you're not interested in—if they are a background artist or passerby, for example.

5 Select the Person 2 group, right-click the clip, and choose Not Person 2 > Remove.

6 Repeat for Persons 3, 4, and 5.

7 Rename the remaining Person 6 as **Fire Dancer**.

8 Select the Other People group.

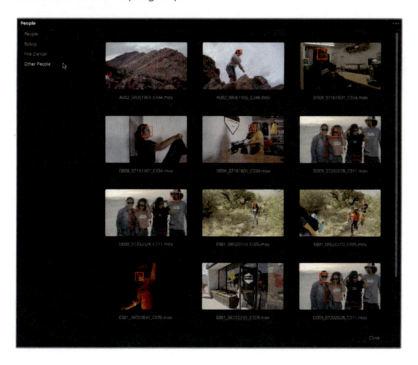

This group contains faces that could not be identified by the analysis, including those you have subsequently removed.

9 Select the first two clips, which are shots of Bobby on the rocks, and then right-click and choose Tag As > Bobby.

NOTE You won't see any additional clips appear in Bobby's group since these clips have already been tagged as Bobby. However, Resolve identified a face in a different part of the clip that it could not identify. Confirming this as Bobby rather than a different person helps prevent misidentification.

10 Right-click the other instance of the fire dancer clip and choose Tag As > Fire Dancer.

You can also add new persons to the People window.

11 Select the two faces of the girl carrying the T-shirt.

12 Right-click them and choose Tag As > New Person.

13 In the Input New Name window, type Kayleigh to name this person, and click OK.

14 Click Close to close the People window.

> **NOTE** To further refine the people metadata, choose Workspace > People. To reset the Face database for the current project, open the People window and choose "Reset Face Database" from the options menu (three dots).

15 In the Smart Bins list, click the disclosure arrow for the People group to reveal the People smart bins.

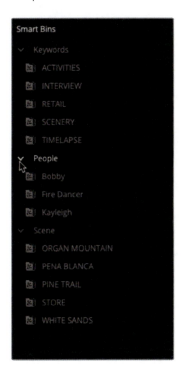

You can also view the People tags associated with each clip in the Metadata panel.

Renaming Clips with Metadata

Clip names from a camera, or almost any capture device, are often an alphanumeric string that typically includes the date and time that the clip was created. They are not always the most descriptive names and often need to be changed for editing purposes. Entering clip names manually is one way to address this, but it is not the only way (or even the most efficient way) to rename them.

Variables are references to other metadata that exist on the clip, such as scene, take, and shot number—so called because variables are not the same for each clip. You can enter a variable into the clip name, and Resolve will reference the correct information for each clip (provided the information is present). You will use the metadata you've imported to change the generic names of the clips in the media pool to more descriptive names.

1   Select the B-ROLL bin and press Command-A (macOS) or Ctrl-A (Windows) to select all the clips in the media pool.

2   Right-click any of the selected clips, choose Clip Attributes, and select the Name tab.

3   In the Clip Name field, type **%** (percentage sign).

Entering % indicates that you are about to enter a variable. When you enter that %, a list of variables appears.

4   Type **sc**.

A list of potential variables appears that contain the letters "sc."

5   In the dropdown menu, click Scene to add it to the Clip Name field.

6   Press the Spacebar to add a space after this variable, and then type **%comm** and choose Comments from the list of variables.

7 Click OK to apply the changes you've just made.

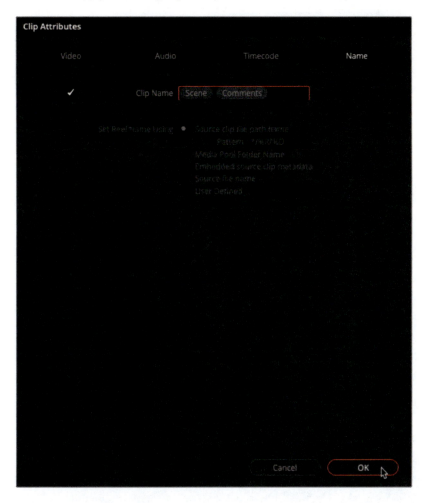

The clip names now show a combination of keywords and descriptions pulled from the clips' various metadata fields.

You can also combine text that you enter along with the variables to create a more descriptive clip name.

8   Select all the clips in the VIDEO bin, right-click them, and choose Clip Attributes.

9   In the Name tab, select the name field and type **LANG %Keywords %CAM # %Take**, selecting the options for Keywords, Cam #, and Take as they appear, and then click OK.

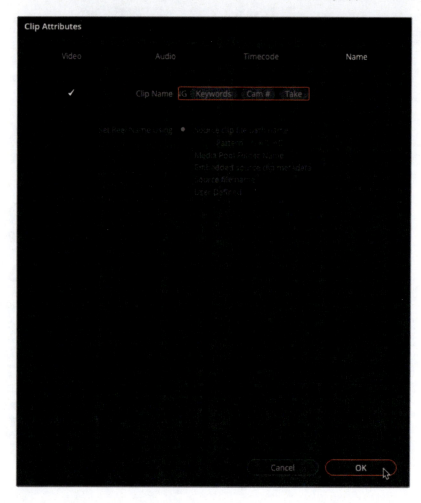

Now all of Chris's interview clips have been renamed with his surname, the clip's keywords, camera #, and take number.

Importing metadata and using it to rename clips with variables can save hours of manual typing and provide clear, descriptive information.

Creating Custom Smart Bins

Using the various automatic Smart Bins options is a great way to help add order to your projects, but the true power of Smart Bins comes to the fore when you can set your own rules for what a Smart Bin will contain. In this exercise, you'll create your own Smart Bins to be able to find and work with certain media from this project.

1   Choose File > New Smart Bin.

The Create Smart Bin window appears. You'll use this smart bin to easily locate all the interview clips from a particular angle.

2   Name this Smart Bin **INTERVIEW A CAM**.

3 Click the second dropdown menu (currently File Name) and choose Keywords.

4 Ensure that the third dropdown menu is set to "contains" and type **interview** in the final field.

This means that any clip that has the "interview" keyword applied will be currently included in this Smart Bin.

5 Click the Add Criteria (plus) button to add another set of criteria.

6 Change the first dropdown menu from "Media Pool Properties" to "Metadata – Shot & Scene."

7   Change the second dropdown menu to "Camera #," the third to "is," and type **A** in the final field.

> **TIP** To navigate quickly through the list of options, type the first letter of the metadata field you need. This will move to the next option in the list starting with that letter. You can then select it from the list with your mouse. Keep pressing the same letter to jump to the next option starting with that letter.

This sorts the contents of this smart bin to include only interview clips from the "A" camera.

8   Click Create.

The new smart bin is added to the Smart Bins list.

An alternative is to create a smart bin for a particular set of clips that you might want to use for a multicam edit.

9   Right-click the INTERVIEW A CAM smart bin and choose Duplicate.

10 Change the name of the duplicated smart bin to **INTERVIEW TAKE 2**.

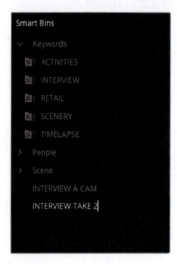

11 Right-click the INTERVIEW TAKE 2 smart bin and choose Edit.

12 In the Edit Smart Bin window, change the "Camera #" option to "Take" and ensure that the third menu is set to "contains."

13 In the final field, type **2** and click OK.

The smart bin now contains the three clips for the second take from the interview.

To organize your smart bins, you can add them to folders, similar to how the Keyword and Scene Smart Bins are grouped.

14 Right-click in an empty space of the smart bins list and choose Add Folder to create a new folder named "Folder 1."

15 Highlight the name for Folder 1 and type **My Smart Bins**.

16 Shift-click the INTERVIEW A CAM and INTERVIEW TAKE 2 smart bins and drag them into the My Smart Bins folder.

As you can see, coupled with an understanding of metadata, Resolve has some flexible and powerful searching functions, so you should always be confident that you'll be able to find your media. One word of caution, however, is that metadata searches are only as good as the quality of the metadata provided in the first place. Sometimes a simple spelling mistake can thwart all these potential benefits.

> **TIP** To make a Smart Bin available across different projects, click the Show In All Projects checkbox. While this doesn't make the contents of the Smart Bin available in other projects (see the "Power Bins" section later in this lesson), it will enable you to reuse the rules for that Smart Bin in another project (including old projects). This is very useful if you often use the same rules for Smart Bins across multiple projects. These Smart Bins are available in a Smart Bin folder called "User Smart Bins."

Using Additional Match Options

Custom smart bins can be created using the options to Match using All or Any of the listed rules you choose. However, for added flexibility you can choose to add additional match options by Option-clicking (macOS) or Alt-clicking (Windows) the Add Filter Criteria button. This will add a new subset of rules that have their own All/Any options.

For example, in the preceding image, the smart bin is set to include any media pool content that has a resolution of 3840 x 2160. However, because using only this value may include unwanted items that could have the same resolution (such as timelines or multicam clips), an additional set of rules has been added, set to Any, that specifies the contents should be video clips, video clips with audio, or still images that match that resolution.

Creating Subclips

Another technique commonly used by editors dealing with large amounts of footage, especially long clips, is to create a series of subclips. You encountered subclips in Lesson 1 when you edited soundbites from the interview, and the subclips made it easier to work with smaller portions of footage rather than the very long interview clips they were taken from.

The important thing to remember about subclips is that, while they are created from a longer clip, they refer to the same source media files on your system. As a result, they don't take up any more space on your system, irrespective of how many subclips you create.

Also, while subclips will initially inherit the metadata of the clip they are created from, the subclip itself is a completely independent clip. This means that you can store them in different bins, and they can each have their own metadata.

You will create a series of subclips from the main interview camera you used in Lesson 1.

> **NOTE** In this exercise, you'll create subclips in the media page. However, the same techniques can be used to create subclips from clips in the source viewer on the edit page.

1 In the INTERVIEW bin, create a new bin and name it **SUBCLIPS**.

2 Select all the clips in the INTERVIEW A CAM smart bin and drag them into the viewer.

3 Select the SUBCLIPS bin.

4 Ensure that LANG INTERVIEW A 002 is active in the viewer, click the viewer's options menu, and choose Show Full Clip Audio Waveform.

5 Set an In point just before Chris says, "My name's Chris Lang…" and an Out point after he says, "…Las Cruces, New Mexico."

> **TIP** You don't need to be accurate when setting In and Out points for subclips. In fact, setting them to include a little more than you intend to use is a recommended technique (see the sidebar "Adjusting Subclip Limits" later in this lesson).

6 Choose Mark > Create Subclip or press Option-B (macOS) or Alt-B (Windows).

The New Subclip window opens, asking you to confirm the name of this subclip, which is the same filename as the original clip but with the word "Subclip" added to the end.

7 Click Create to add the subclip to the current bin.

8 From the viewer dropdown menu, choose the clip **LANG INTERVIEW A 005** and set an In point just before Chris says, "That's what really inspires us…" and an Out point after he says "…the design process starts."

9 Right-click the scrub bar and choose Create Subclip.

10 Click Create in the New Subclip window.

11 From the viewer dropdown menu, choose the clip **LANG INTERVIEW A 008**.

12 Set an In point just before Chris says, "We want people to experience..." and an Out point after he says "...the landscapes."

13 Drag the clip from the viewer into the SUBCLIPS bin.

14 Again, click Create in the New Subclip window.

15 Continue playing the clip in the viewer and set another In and Out point around Chris's final soundbite: "That's why we say experience the southwest."

16 Choose your preferred method for creating a subclip from this last soundbite and click Create in the New Subclip window.

You now have four subclips in the SUBCLIPS bin.

Modifying Subclip Metadata

Now that you've created the subclips, you can take advantage of some of the metadata tricks you learned throughout this lesson.

1   Select the first subclip and, in the Comments field in the Metadata panel, type **intro**.

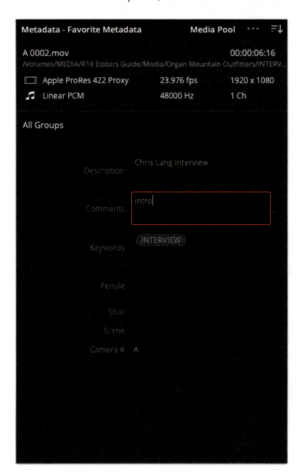

2   Select the second subclip and, in the Comments field, type **inspires us**.

3   Select the third subclip and type **so unique** in the Comments field.

4   Select the final subclip and type **tagline** in the Comments field.

5 Select all the clips in the SUBCLIPS bin and, in the Metadata panel, delete the INTERVIEW keyword and type **SUBCLIP** the Keyword field.

6 Click Save at the bottom of the Metadata panel.

7 Right-click the subclips and choose Clip Attributes.

8 In the Clip Name field, type **LANG %Keyword - %Comments**, selecting the metadata options from the menus as they appear.

9 Click OK to rename the subclips.

You can now edit using these subclips just like any other clip, as you did in Lesson 1.

Adjusting Subclip Limits

One limitation of a subclip is that, even though it is referencing the original source media file, it is limited to the initial In and Out points you used to create it. Therefore, it's good practice to set your initial In and Out points a little before and after the portion you want to subclip, thereby leaving a little bit of wiggle room when you later trim the clips in the timeline.

However, when you find that you need a few extra frames not included in your subclip, you can always extend the limits of a subclip by right-clicking a subclip and choosing Edit Subclip.

This allows you to adjust the start and end timecodes for the subclip, updating the limits of the subclip in both the media pool and the timeline simultaneously.

Power Bins

Throughout this lesson, you've been using both regular bins and Smart Bins, both of which are great organizational tools to use within a project. However, a third bin type available in DaVinci Resolve is the Power Bin. Unlike regular bins or Smart Bins that exist only within the current project, Power Bins are available across every project within the current Project Library. They are useful for storing elements you want to reuse across separate projects, such as graphics, titles, sound effects, or music files.

1   Click the media pool's options menu and choose Show Power Bins.

Power Bins are displayed in the media pool above the Smart Bins.

2   Select the Power Bin's Master bin, and press Shift-Command-N (macOS) or Shift-Ctrl-N (Windows) to add a new Power Bin. Name the bin **LOGOS**.

3   Select the Graphics bin that's currently in your project, which contains a graphic file called OMO LOGO.png.

4   Drag this clip to the LOGOS Power Bin.

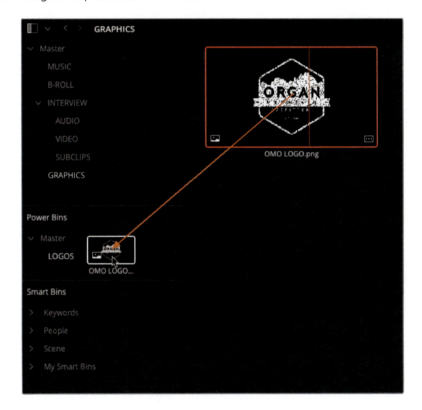

Power Bins are accessible from every project in the current Project Library, so this image file (plus its associated metadata) will now be easily accessible for any other project. The next time you find yourself with a series of projects that share elements, whether they be sound effects, graphics, or common video elements, you can use Power Bins to save time manually importing clips and any required metadata into different projects.

NOTE Multicam clips, compound clips, timelines, and Fusion clips cannot be placed in Power Bins.

You should now have a good understanding of just how powerful Resolve is at helping you organize your projects. Remember, while some projects may require much more organization than others, the techniques demonstrated throughout this lesson can be applied at any point you're working on a project and in many cases can be applied as readily in the edit page as they can in the media page. Project organization doesn't just occur once, at the start of a project; it is something that you will constantly refine as you are editing.

Lesson Review

1   True or false? You can save a preset of your current project settings to make future project configuration easier.

2   Which application can be used to easily create proxy media for editing?

    a)   Resolve Proxy Maker

    b)   Blackmagic Proxy Generator

    c)   Blackmagic Proxy Creator

3   True or false? Proxy media replaces the source media files used in your project.

4   What method(s) can you use to auto sync sound to video files in the media page?

    a)   Waveform

    b)   Timecode

    c)   Markers

5   Which type of bin allows you access to its contents across different projects in the current project library?

    a)   Smart Bins

    b)   Super Bins

    c)   Power Bins

Answers

1 True. Project presets are saved in the Presets panel of the Project Settings.

2 b) Blackmagic Proxy Generator.

3 False. Proxy media files are created alongside your source media files. Your original, full-resolution media files remain intact. Proxy media can be disabled by choosing Playback > Proxy Handling > Disable All Proxies.

4 a) Waveform and b) Timecode.

5 c) Power Bins.

Edit Page Effects

Building graphics sequences is a valuable skill in any editor's arsenal. Multilayered composites that combine video, audio, graphics, and text, often with animated elements, are regularly incorporated into an edit. And it often falls to the editor to build these composites and apply some level of keyframed animation, either as pre-viz (previsualization) for motion graphics or as the final content.

This lesson will introduce the compositing and animation features available to you in DaVinci Resolve 18's edit page, together with some specialized effects to help you achieve common tasks quickly and effectively.

Time

This lesson takes approximately 60 minutes to complete.

Goals

Setting Up the Project

To begin this lesson, you will import a DaVinci Resolve project and configure the edit page workspace for working with effects.

1   Open DaVinci Resolve, and in the Project Manager, right-click an empty area and choose Import Project. Navigate to R18 Editors Guide/Lesson 07.

2   Select the **EDIT PAGE EFFECTS.drp** project file and choose Open.

The Project is imported and added to the Project Manager.

3   Open the project, click the Edit page button, and relink the media files.

4   If necessary, choose Workspace > Reset UI Layout to reset the interface to its default setting.

This project contains several different bins with timelines for the different exercises you'll follow throughout this lesson.

5   Click the Select Timeline pop-up menu above the timeline viewer.

The Timeline pop-up menu lists all timelines in the current project and can be a useful way of easily switching timelines. By default, Resolve arranges this list of timelines in Recently Used order, so recently accessed timelines appear at the top of the list.

6   Click the timeline viewer Options menu and choose Timeline Sort Order > Alphabetical.

7   Click the Select Timeline pop-up again to view the sort order change and ensure that the 01 Title Composite timeline is currently open.

> TIP You can also access the timeline sort order options in DaVinci Resolve > Preferences > User > UI Settings and change the Timeline Sort Order pop-up menu.

You're now ready to start the first exercise in this lesson.

Compositing Using Traveling Mattes

A common task that many editors face is the need to resize clips. This could be as simple a changing the framing of a shot to exclude a microphone that's unintentionally wandered into the frame, or it could be to build something a little more creative.

In the first exercise in this lesson, you'll composite a shot with a traveling matte and then resize and position the result into the overall composite.

The 01 Title Composite timeline is a short sequence of archive footage composited over a background image. Each of the archive clips has a stylized "painted" edge to it. After a second or so, an animated Fusion title animates in, announcing the title of the film: Living In The Age Of Airplanes.

Your job is to use the compositing functions available in the edit page to stylize the final archive image and place it in the bottom-right corner of the timeline viewer.

First, you'll need to create a new video track in the timeline to accommodate the new clip.

1 In the bin list, select the Archive bin and open the clip AOA Archive 4 in the source viewer.

This clip is a short archive clip from the early days of powered flight.

2 Right-click any of the track headers in the timeline and choose Add Tracks.

The add tracks window appears. You want to add a new track above the current V4 but below V5.

3 Leave the Number of Video Tracks set to 1 and change the Insert Position to Above Archive 3. Change the Number of Audio Tracks to 0 and click Add Tracks.

A new, empty V5 track is created in the timeline.

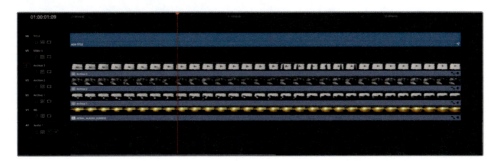

4   Rename the new V5 track **Archive 4**.

You now need to mark a duration for the new clip and target the new track.

5   Select the clip on the track below the track you have just added and press Shift-A to mark the duration of the selected clip.

6   In the track header, click the destination control for V5 to target the new track.

You have currently marked a duration of 5 seconds in the timeline, but you only have just over 3 seconds of the source clip.

For the source clip to correctly fill the marked duration in the timeline, you'll use a Fit to Fill edit.

The Fit to Fill edit is the final type of edit you can perform in DaVinci Resolve's edit page and is another four-point edit. Fit to Fill uses the duration of the In and Out points in the timeline and source viewers and adjusts the speed of the source clip to match.

7   Drag the clip from the source viewer to the Fit to Fill option in the timeline viewer overlays.

The clip is edited on to the targeted track, and its speed is automatically adjusted as indicated by the speed icon next to the clip's name in the timeline.

8   To check the new speed of the clip, select the AOA Archive 4 clip in the timeline, open the Inspector, and reveal the Speed Change controls.

The clip is now running at about 65% of its original speed.

9   Close the Inspector.

NOTE When editing a longer source clip to a shorter timeline duration, the speed of the source clip is increased proportionately with the shorter duration.

Creating a Compound Clip

To create the "painted edge" effect to this clip, you need to combine the clip with a matte clip on another timeline track. However, to keep this timeline as neat as possible and not to add tracks unnecessarily, you will nest the clip in its own compound clip.

> **NOTE** You briefly encountered compound clips in Lesson 3 when editing footage from one timeline to another. Compound clips are a useful way of grouping clips together into one timeline element, much like a multicam clip, and are often used in compositing tasks such as this to achieve certain results.

1 In the media pool, select the Compound Clips bin.

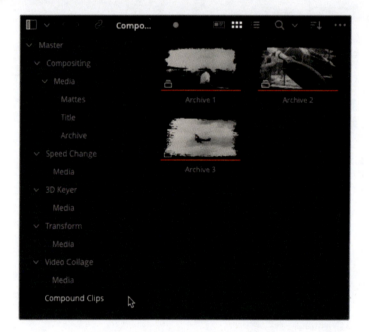

This bin already includes several compound clips that were already created for this project.

2 In the timeline, ensure that the **AOA Archive 4** clip is still selected and choose Clip > New Compound Clip.

The New Compound Clip window opens.

3 Name the compound clip you are creating **Archive 4** and click Create.

The selected clip is now nested into the new Archive 4 compound clip, as indicated by the compound clip icon next to the compound clip's name in the timeline.

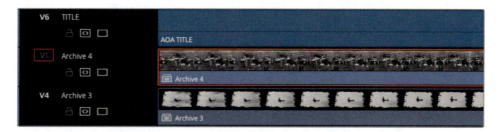

The new compound clip has also been added to the selected bin in the media pool.

Adding the Traveling Matte

Next, you'll create the stylized edging for this clip. To do so, you'll add an image file to use as a matte inside the compound clip.

1   In the Timeline View Options menu, enable Stacked Timelines and display the timeline tabs.

2   Ensure that the **Archive 4** compound clip is still selected in the timeline and choose Clip > Open in Timeline.

The compound clip opens in its own timeline tab, much like the multicam did in Lesson 5. Similar to that multicam clip, any changes you make in this timeline will be seen in the compound clip in the main **01 Title Composite** timeline.

3 Select the **AOA Archive 4** clip in the compound clip timeline and move it up to automatically create a new V2 track.

4 With the **AOA Archive 4** clip still selected, press Shift-A to mark its duration.

5 From the Mattes bin in the media pool, open the **matte 4 alpha.png** clip in the source viewer.

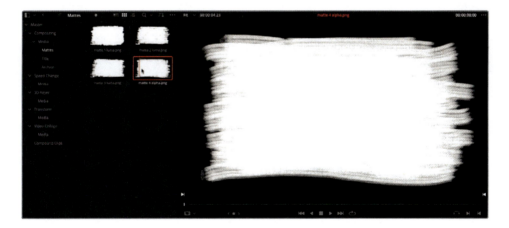

This image was created using a graphics package and was saved as a PNG file with an alpha channel that you will use to composite it with the **AOA Archive 4** clip.

6 Drag the matte from the source viewer to the Fit to Fill overlay, or press Shift-F11, to edit the clip onto the currently targeted track in the timeline, track V1, at the correct duration.

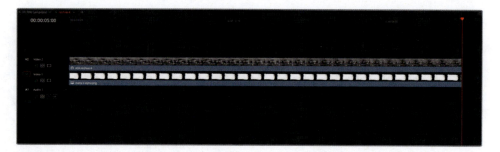

You now need to tell Resolve how these two layers should interact with each other in the timeline.

7 Select the **matte 4 alpha.png** clip in the timeline, open the Inspector and, in the Composite controls, change the Composite Mode pop-up menu to Alpha.

8 Select the **AOA Archive 4** clip in the timeline and change the Composite Mode to Foreground.

The Foreground clip is now composited with the matte clip using the lower clip's alpha channel.

9   Close the Archive 4 timeline tab to close the compound clip's timeline and return to the 01 Title Composite timeline.

You can now see that the clip is composited within the main 01 Title Composite timeline.

Resizing the Composited Image

The final stage is to adjust the size and positioning of this clip in the overall composite.

1 Select the **Archive 4** compound clip in the timeline and choose View > Viewer Overlay > Transform to enable the onscreen controls.

2 Use the corner handles to resize the selected image to about a third of its original size.

> **TIP** You can use the Zoom controls in the Video tab in the Inspector to specify precise values if required.

3 Drag the image into the lower-right corner of the timeline viewer.

Finally, you will add a slight sense of perspective that's been added to all the previously composited clips.

4   In the Inspector, adjust the Yaw of the selected clip to about -0.05 to add a slight "twist" to the image in the timeline viewer.

5   When you've made your adjustments, choose View > Viewer Overlays > Toggle On/Off or click the timeline viewer's overlay control to turn off the onscreen controls and play back the composited titles to review your work.

Variable Speed Changes

Making changes to the speed at which a clip plays back in the timeline is a common task that can be easily achieved in the Speed Change controls in the Inspector. However, many editors are often called upon to make more creative speed change adjustments that often involve changing the speed of a clip over time, so that one part of a clip plays at a different speed than another part. This technique is commonly referred to as variable speed changes, which can be achieved using Resolve's Retime controls.

1   From the Timeline pop-up menu, choose the 02 Speed Changes timeline.

2   From the top of the timeline viewer, select Single Viewer Mode.

The current timeline contains two clips to which you will apply different speed changes. The first clip, DISPLAYING SHIRT, is a long clip that follows the journey of an Organ Mountain Outfitter's T-shirt on its journey from having been printed in the backroom to being displayed and ready to buy in the shop.

Currently, the clip is running at around 40 seconds from start to end. You can make this much more visually interesting, and also reduce the duration of the shot, by adjusting the speed of the clip.

3   In the Inspector's Speed Change controls, select the Ripple Timeline option and then change the Speed% to **800**.

Because you changed the Ripple Timeline option before you adjusted the speed of the clip, the rest of the footage in the timeline has rippled to accommodate the clip's new, retimed duration.

Adding the Speed Points

To create a variable speed change, you need to add one or more speed points.

1   In the timeline, place the playhead somewhere around the point where the first girl is passing the T-shirt to the second girl (around a third of the way into the clip).

2   Select the clip in the timeline and choose Clip > Retime Controls or press Command-R (macOS) or Ctrl-R (Windows) to display the clip's retime controls in the timeline.

3   Click the Clip Speed pop-up menu (the black triangle next to the clip's speed) and choose Add Speed Point.

A Speed Point is added to the clip at the location of the playhead, and the retime controls now display two 800% speed segments on this clip.

4 Move the playhead forward until the second girl has taken the T-shirt and is reaching for the clothing hanger.

5 Click the Clips Speed pop-up menu for the second speed segment and choose Add Speed Point to add an additional speed point at the playhead's location in the clip, thereby creating a third speed segment.

6 Add a third speed point where the second girl hangs the T-shirt in the shop.

By adding three speed points, you have created a total of four distinct speed segments on this clip. Each of these segments can have their own speed settings.

7   Press T to enable Trim Edit mode.

With Trim Edit mode enabled, any adjustments you make affecting the clip's overall duration will ripple the timeline.

8   Click the Clip Speed pop-up menu for the second speed segment (after the first speed point) and choose Change Speed > 400%.

This segment of the clip, between the first and second speed points, will now play back at 400% of the clip's original speed.

9   Click the Clip Speed pop-up menu for the fourth speed segment (after the third speed point) and choose Change Speed > 200% to adjust that part of the clip.

You can also manually adjust the speed of the individual segments.

10 Select the top part of the second speed point and drag to the right to decrease the speed of the preceding speed segment to about 300%.

NOTE Manually adjusting the control points like this always affects the speed of the segment of the clip to the left of the speed point. To adjust the speed of the final speed segment of a clip (after the final speed point), select and drag the right edge of the Change Speed name bar at the top of the clip.

You can also refine the position of each speed point, adjusting the starting frame for each speed segment.

11 Select the lower part of the first speed point and drag to the left until the viewer shows the girl behind the door to the store.

This changes the location of this speed point (between the first and second speed segments) to an earlier frame.

12 Drag the same control to the right until the girl is just about to hand the T-shirt over to the second girl.

Again, this refines the position of the speed point to a slightly later frame.

Adjusting the Retime Curve

You can make further changes to the speed of the segments between the speed points, as well as the interpolation between each of the speed sections, using the clip's Retime Curve.

1   Right-click the **DISPLAYING SHIRT** clip in the timeline and choose Retime Curve to display the retime curve below the clip in the timeline.

You can use two retime curves to finesse the speed of the clip. By default, the Retime Frame curve appears below the timeline clip. A diagonal line represents the speeds at which the various parts of the clip progress, with steeper parts of the curve representing faster speed sections. Moving any of the control points horizontally changes the location of the speed point within the clip. Moving any of the control points vertically will change the speed of the preceding and subsequent segments: moving a speed point up will increase the speed of the preceding speed segment and decrease the speed of the subsequent segment; moving the control point lower will slow the preceding segment and speed up the subsequent segment. The overall result will not adjust the duration of the clip and is useful if you don't want the clip to change duration in the timeline.

2   Click the Curve pop-up menu in the top left of the retime curve.

3 In the curve pop-up menu, uncheck Retime Frame and check Retime Speed.

This changes the curve to display the speed as a horizontal line, with the relative speeds of the segments indicated depending on how high the bar is on the curve: segments with a higher line on the curve are playing back faster than segments at a lower point.

4 Drag the horizonal curve between the second and third speed points up to around 900%.

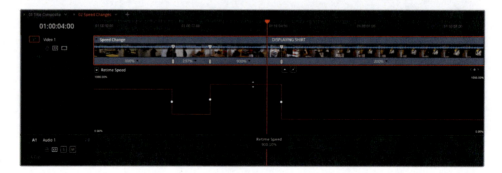

5 Click to select the first speed point in the Retime Curve.

6 Click the Bézier Curve button to add Bézier handles to the selected speed point.

Adding the Bézier handles changes the way in which the different speeds of the two segments are interpolated. Now, instead of an instant change from one speed to another, the change happens gradually over several frames.

7 Drag the Bézier handles further apart to make the interpolation more gradual or drag them closer together to make the change more abrupt.

8 Select the other two speed points in the Retime Curve in turn and add Bézier handles to each, adjusting the interpolation between each speed segment to your liking.

> **NOTE** To close the Retime Curve when you're finished, choose Clip > Show Curve Editor or press Shift-C.

9 When you've finished making the adjustments, right-click the clip and choose Retime Curve to close the retime curve and press Command-R (macOS) or Ctrl-R (Windows) to close the clip's Retime Controls.

The different speed controls can be used either in conjunction with each other or independently, depending on your needs. For example, you might want to add and adjust the speed points using the intuitive Retime Controls before refining their interpolation using the Retime Speed curve. Alternatively, you might want to add and adjust the speed points for a clip using only the Retime Frame curve. The choice is entirely yours.

Creating Freeze Frames

Another common requirement when dealing with speed changes of a clip is to have the clip slow to a freeze frame. This technique is just as frequently used as a dramatic introduction for a character in a film, or as a teaching aid as part of a training video.

1 Zoom and scroll the timeline to the next clip, called ST MAARTEN OVERHEAD.

2 Place the playhead on a frame where the undercarriage of the airplane is clearly visible.

3   Select the **ST MAARTEN OVERHEAD** clip in the timeline and press Command-R (macOS) or Ctrl-R (Windows) to display the Retime Controls for the clip.

4   Click the Clip Speed pop-up and choose Freeze Frame.

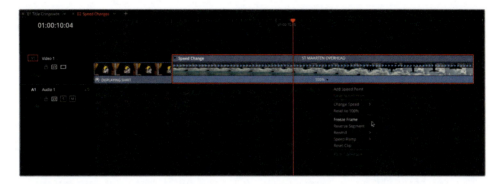

Two Speed Points are added to the clip, with a 2-second freeze frame between them.

5   Right-click the clip **ST MAARTEN OVERHEAD** and choose Retime Curve.

6   In the Curve pop-up menu, deselect the Retime Frame curve and select the Retime Speed curve.

7   Add Bézier handles to each of the speed points in the Retime Speed curve.

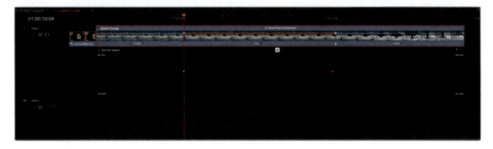

8   Play back the **ST MAARTEN OVERHEAD** clip to review the interpolation from 100% to 0% speed and back to 100% again.

Changing Retime Processing and Motion Estimation

You'll no doubt notice that the change between the speed segments isn't quite as smooth as you'd probably like. This is due to the Retime Process used by the clip.

1   At the bottom of the Inspector, reveal the Retime and Scaling parameters.

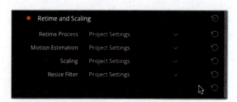

Currently, the clip is using Project Settings for any retime processes you apply to it, which, for this project, is set to the Resolve default of Nearest. This option essentially repeats video frames as necessary when slowing a clip's speed below its original frame rate and accounts for the somewhat stuttering playback of this clip as it approaches and leaves the freeze frame. Two other options are available in this menu: Frame Blend and Optical Flow.

2   Change the Retime Process to Frame Blend and play back the clip to review the change.

Frame Blend applies short dissolves between the duplicated frames in an attempt to smooth out the movement within the slowed clip. Depending on the footage, it can often create a smoother result than using Nearest.

3   Change the Retime Process option to Optical Flow and, again, review the change.

> **TIP** For less powerful computers, you may need to choose Playback > Render Cache > Smart for the next few steps to allow Resolve to cache the speed changes in the timeline since Optical Flow processing is a very processor-intensive computation.

Optical Flow is a high-quality but processor-intensive process that uses motion estimation to generate new frames from the original source frames. It can often produce vastly superior results when playing back a clip at speeds slower than its original frame rate as long as you don't have one object passing in front of another within the frame. If this is the case, you may notice "rippling" artifacts around the edges of objects, which to a degree are still visible in this clip.

DaVinci Resolve, however, provides further settings to help minimize this problem by allowing you to adjust the Motion Estimation being used by the Optical Flow process.

> **NOTE** Motion Estimation will only be applied if the Retime Process is set to Optical Flow in the clip or Project Settings.

Again, Motion Estimation has several settings that become increasingly demanding: Standard Faster and Standard Better are processor-efficient processes that often yield good-quality results for most situations. The default for Project Settings is Standard Faster.

Enhanced Faster and Enhanced Better should yield superior results in cases where the standard options exhibit artifacts. Both of these options are more computationally intensive that the Standard options, so they take longer to process.

Finally, Speed Warp allows for even better results when reducing the speed of a clip below its original frame rate. Speed Warp uses the DaVinci Resolve Neural Engine to reduce visual artifacts to a minimum. Due to its high processor demands, it is only available on a clip-by-clip basis and cannot be set as an option in the Project Settings.

As always, results for each setting will vary depending on your footage.

4   Change the Motion Estimation setting for the ST MAARTEN OVERHEAD clip to Enhanced Better.

Even at this setting, you should notice a slight problem with the way the wing tips are being interpolated.

5   Choose Playback > Render Cache > User.

6   Right-click the **ST MAARTEN OVERHEAD** clip in the timeline and choose Render Cache Color Output.

> **NOTE** Render Cache format settings and other controls are available under Optimized Media and Render Cache in the Master Settings category of the Project Settings window.

7   Change the Motion Estimation setting to Speed Warp.

8   Allow the Speed Warp process to complete and the timeline to automatically cache the results. Once completed, play back the results.

With Speed Warp enabled in the Motion estimation settings for this clip, the results are vastly superior. Speed Warp, however, is such an intensive process that it should only be used sparingly on the most high-spec machines.

> **NOTE** If you're using the free version of DaVinci Resolve, any clip using the Speed Warp motion estimation setting will be watermarked.

Render In Place

The edit page has an additional feature for assisting you if you're dealing with processor-intensive effects, such as this Speed Warped clip. Render in Place allows you to easily render out selected timeline clips to an entirely new media file on your hard drive, together with any "baked in" effects. The rendered media is then automatically added to the media pool and used to "replace" the original timeline clip. This is useful when you need to grade a clip with multiple effects since it means you don't have to keep caching that portion of the timeline.

> **NOTE** Render in Place can be applied to multiple selected timeline and compound clips. Selecting multiple clips will result in each clip being rendered in place separately, but as part of a batch operation.

1   In the media pool, select the Speed Change bin and, in the timeline, right-click the ST MAARTEN OVERHEAD clip and choose Render in Place.

The Render in Place window opens displaying the options that will be used for creating the new media files.

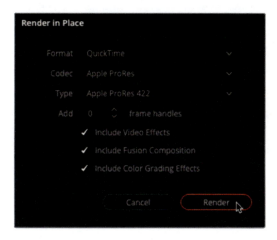

2   Leave the options set to their defaults and click Render.

You'll be asked where the new media should be placed on your system.

3   Navigate to R18 Editors Guide/Lesson 07, select the Render in Place folder, and click Open.

The new media file is immediately created in the chosen location.

Once the process is complete, the original **ST MAARTEN OVERHEAD** timeline clip has been replaced with the newly created file called **ST MAARTEN OVERHEAD Render 1.mov**, which is also added to the selected bin in the media pool.

Because this clip now has the speed ramp "baked in," it will now play without the need to cache the clip in the timeline. However, the speed changes you've made are no longer accessible.

Thankfully, Render in Place is not just a one-way operation. Once a clip has been rendered in place, you can always revert the process if you need to refine the effects further.

4   In the timeline, right-click the **ST MAARTEN OVERHEAD Render 1.mov** clip and choose Decompose to Original.

The original clip, along with the editable speed changes, is returned to the timeline but the rendered version remains in the media pool and on your hard drive. This allows you to refine the effect further before creating an additional rendered in place clip, if required.

> **TIP** It can be useful to color code clips as you render them in place so that you can quickly identify which clips can be decomposed to original later. Alternatively, you can simply place the new clip on a video track above the original clip in the timeline.

3D Keyer FX

A common effects task many editors are faced with is having to work with green- or blue-screen footage. This footage is designed to be *keyed* against a background plate and is a common technique used in fantasy and science fiction films, as well as in many TV shows and online videos, where you may want your talent to appear against a virtual background.

Alternatively, even if the shot will be finalized later by a dedicated visual effects artist, the editor may still need to create a temporary version of the composite to see how effectively the foreground and background elements will work together. If the shots can't be readily combined as planned, it is so much easier for the editor to swap takes or edit around the identified problems than it would be for the VFX artist.

In this exercise, you'll use the new 3D Keyer effect to composite a shot from the short sci-fi film *Hyperlight*.

1   From the timeline viewer pop-up menu, choose the **03 3D Keyer** timeline.

2   Play the timeline to review the short scene that's been edited for you, and then return your playhead to the start of the clip on Video 2.

This is a short sequence of shots from the sci-fi short Hyperlight. Obviously, the second shot of the man against the green screen somewhat ruins the conceit that these two characters are supposed to be in a spaceship orbiting a planet.

3 Select the clip on Video 2 and press D to disable the clip in the timeline and reveal the background plate of the planet.

This is the view that the man should have outside the window of the spaceship. To achieve this, you need to apply the 3D Keyer filter.

4 Press D again to re-enable the clip in the timeline.

5 Click the Effects button to open the Effects Library, select the Open FX > Filters group, and scroll through the list of filter categories to locate the Resolve FX Key filters.

Each of the Resolve FX Key filters is based on the appropriate color page qualifiers; indeed, these filters are effectively the respective controls from the color page packaged and made available in the edit page. There are two color key filters you can choose to work with. For this exercise, you'll use the 3D Keyer but, even though the controls are broadly the same, the HSL Keyer is useful if you want to specifically target different combinations of hue, saturation, and brightness for a more refined key. The Luma Keyer is used for performing keys on just the luminance (brightness) of a clip and is often used in conjunction with composite modes. The Alpha Matte Shrink and Grow filter can be used to further refine the keys created using the other filters.

6 Double-click the 3D Keyer filter to apply it to the selected green-screen clip of the man on Video 2.

The only indications that the effect has been applied is the appearance of the FX badge in the timeline clip's information and the Effects tab of the Inspector being highlighted.

7 Select the Effects tab in the Inspector to reveal the control for the 3D Keyer filter.

To begin the process of compositing this shot with the background plate, you'll need to enable the viewer's onscreen effect controls.

8 Choose View > Viewer Overlay > Open FX Overlay.

This enables the onscreen controls for any effects that have controls you can access in the viewer.

9 With the first Eyedropper tool selected in the 3D Keyer controls, click and drag across the center area of the green screen.

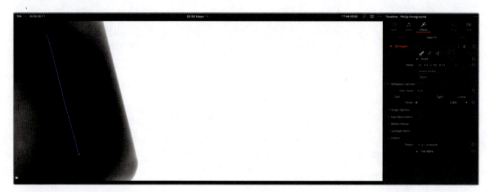

This process samples the main hue, saturation, and luminance values of the green screen highlighted by the blue onscreen control. However, you'll notice that some parts of the green have fallen outside of this selection.

10 Click the disclosure arrow for the Usage Options and select the Show Paths option to display the selection path you have chosen.

> **NOTE** Selecting Show Paths automatically deselects the Smart Show Paths option.

11 In the Effects Inspector, click the Add Eyedropper control to add additional areas of the green screen that fell outside of your initial selection.

12 Click and drag again to create additional selections for the key across the areas of green not included in your initial selection.

TIP If you find that you have been too aggressive in your selection and parts of the shot outside the green have been selected, you can attempt to remove those areas from the selection by using the Subtract Eyedropper control.

With additional parts of the green screen added to your selection, you'll probably notice that a green edge remains near the side of the window. To refine the key further, you'll want to see a bit more detail as to what you are selecting.

13 Deselect the Show Paths option in the Usage Options.

14 At the bottom of the 3D Keyer controls in the Effects Inspector, change the Output pop-up menu to Alpha Highlight B/W.

In this view, the alpha channel you're creating for this clip is displayed. White represents a solid area where you won't see the background plate, black represents areas of transparency where the background plate is fully visible, and any areas of gray represent areas of partial transparency.

15 Change the Output pop-up menu to Alpha Highlight.

Alpha Highlight displays the selected part of the key as a flat, gray area and enables you to easily identify the areas of green that remain around the edges of the window.

To refine the key, you need to access the Matte Finesse controls.

16 In the Effects Inspector, click the disclosure arrow to collapse the Usage Options controls, and then click the disclosure arrow for the Matte Finesse controls.

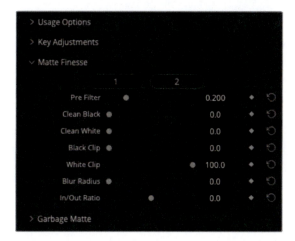

The Matte Finesse controls contain familiar parameters for adjusting the cleanliness of the key. In this case, you just want to clean up the black areas of the key.

17 Increase the Clean Black slider to about 18.0 to further refine the key and include the final, wayward portions of the green screen in your selection.

> **TIP** Use the scroll wheel on your mouse to zoom in and the middle mouse button to pan around the clip in the timeline viewer to get a better look at the matte refinements you're making. Press Z to fit the whole of the shot back in the timeline viewer when you're happy with your results.

18 Change the Output pop-up menu back to Final Composite to see the results so far.

Adjusting Despill

So far, it seems that you have achieved an acceptable result, and the director is at least happy that the foreground and background elements work together as planned. However, as always, there are ways this composite can be improved to help sell the shot further, mainly by removing some of the spill from the green screen.

> **NOTE** As most keying filters rely on selecting a range of hue, saturation, and luminance values to create the key, the process of selecting the correct HSL values can be made slightly easier by having the green screen flooded with light when it is filmed. Unfortunately, depending on the rest of the shot's proximity to the green screen, this can result in unwanted reflection, or spill, of the green on the foreground elements.

1   Play through the clip on Video 2 until the man turns away from the window.

Look carefully and you should see that the man's face is reflecting some of the green. Moreover, the upper-right portion of the man's uniform (where the USEF logo is) isn't supposed to be that green because it also is reflecting light from the green-screen background.

2   In the Effects Inspector, increase the Despill slider in the Keyer Options controls to reduce the amount of spill falling onto the man's face and uniform (you might have to be quite aggressive for this particular example).

The Despill control quickly eliminates the spill from the image while retaining the image's original color.

Excellent job! You've now seen how easy it is to create a composited shot directly in the edit page using the 3D Keyer filter.

Transform FX

Another common compositing task many editors need to undertake is adding elements to enhance shots such as logos on the side of trucks or placing content on screens to make them more relevant to the scene. Earlier in this lesson, you quickly scaled and positioned a clip for a picture-in-picture effect, applying a small change to the yaw parameters for a slight change to the perspective of the clip. However, in many cases you'll need more control over the clip than the pitch and yaw can easily provide. In those cases, you'll need to use the Transform filter to be able to *corner pin* the image into place.

You'll explore this technique by adding a logo to a shop window.

1   From the timeline viewer pop-up menu, select the **04 Transform** timeline.

This timeline contains three shots. The first shot has a .png version of the "Citizen Chain" logo edited over the top of it.

2 In the timeline, select the **CC-Logo.png** clip and, from the Effects Library, locate the Resolve FX Transform group of filters.

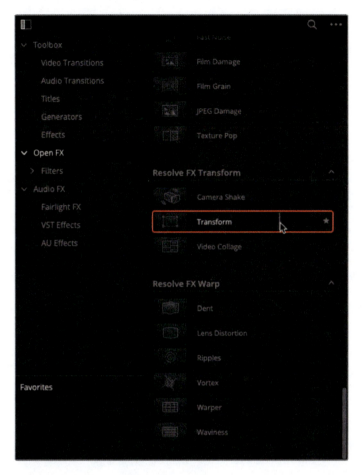

3 Double-click the Transform filter to apply it to the **CC-Logo.png** clip in the timeline and click the Effects tab in the Inspector.

At first glance, you'd be forgiven for thinking that the Transform filter has very similar controls to the standard transform parameters available in the Video Inspector of any clip. However, this filter does add some additional controls over the standard transforms by offering advanced options that include motion blur, edge behavior, and cropping controls.

4   Change the Control Mode pop-up menu to Interactive – Canvas.

The Interactive – Canvas mode moves many of the Transform filter's controls from the Inspector to the timeline viewer, as represented by the white outline and red vertices. You can drag these onscreen controls to distort the image in a variety of ways.

> **NOTE** You may need to choose View > Viewer Overlay > Open FX Overlay if you don't see the onscreen controls you activated in the previous 3D Keyer exercise.

5   Ensure that your playhead is at the start of the timeline and, using the corner areas, drag the controls so that the corners and edges of the graphic align with the corners and edges of the window in the underlying video clip.

6   When you're happy with the way the graphic aligns with the window, start playback.

Oh dear. Now you can see that although the graphic is aligned perfectly, because the camera tracks right as Sasha arrives, the window doesn't stay still within the frame, but the graphic does.

No problem. A couple of keyframes should help.

7   Return your playhead to the start of the timeline and, in the Effects Inspector, click the disclosure arrow to open the Animation controls, and then click the keyframe button for the Canvas Keyframe control.

The keyframe button turns red to indicate that a keyframe has been applied at the start of the clip.

8 Press ' (apostrophe) to move your playhead to the last frame of the **CC-Logo.png** clip in the timeline and adjust the corner areas of the onscreen controls to align the graphic with the window once more.

Another keyframe is automatically added at the end of the clip.

9 Press Shift-` (grave accent) to disable the timeline viewer's onscreen controls and review the results of the keyframing.

Perfect! The graphic now looks like it's part of the shop's window display. The keyframes allow the graphic to follow the movement of the camera. Thankfully, the camera movement is smooth and consistent. If it wasn't, then more keyframes would have been needed.

One final touch will really help sell this effect. You can change the composite mode of the graphic to better integrate it with the background clip.

10 Click the Video tab in the Inspector for the **CC-Logo.png** clip and, in the Composite controls, change the Composite Mode pop-up to Overlay.

The Overlay composite mode increases the contrast of the underlying video clip where the graphic is the brightest (there is no change to the contrast of the clip where the .png is transparent).

11 Reduce the Opacity slider to about 60.0 to reduce the intensity of the graphic and integrate it further with the video clip.

> **NOTE** The final control mode for the Transform filter is Interactive – Pins. Adjusting the image in this mode is done by manually placing control points, called *pins*, in the timeline viewer. Adding one pin only gives you position control. At least two points are required for scaling and rotation. Dragging on one of the pins scales or rotates the image around the other pin. Using three pins, you can create perspective distortions by dragging any one of the pins. You can add up to four pins for unique corner-pinning distortions that don't rely of the regions specified by the Interactive – Canvas mode.

Congratulations! You have now learned how you can distort and keyframe a clip to composite onto other shots. While more comprehensive tools for this sort of task, including trackers, are available in the Fusion and color pages (both of which are outside the scope of this book), knowing how to accomplish a simple version in the edit page is a useful technique for any editor to have up their sleeve!

Video Collage

The final Resolve FX you will work with in this lesson is the Video Collage filter. Previously, you saw how the Transform parameters in the Video Inspector are used to adjust the Zoom and Position of a clip to create a composited effect. The Video Collage filter is designed to make it easier to create uniform, grid-based, picture-in-picture and other split-screen layouts. It is ideal for quickly creating a "video wall" effect, which can require the scaling and positioning of many clips.

The Video Collage filter works in two main ways. The default is Create Background, which you'll explore first.

1 From the timeline viewer pop-up menu, select the **05 Video Collage Background** timeline.

This timeline consists of several video clips stacked on top of each other on different tracks.

Typically, to create a picture-in-picture effect, all the clips on the upper video tracks would need to be scaled and positioned accordingly. However, the Video Collage filter uses the *topmost* clip as the background, using this clip as a "frame" and creating holes in this frame to reveal the clips on the lower tracks—"cookie-cutter" style.

2 From the Effects Library, locate the Resolve FX Transform group of filters and double-click the Video Collage filter to apply it to the clip on Video 4.

The clip on Video 3 is now displayed within four boxes. These four boxes are the holes created in the Video 4 clip by the Video Collage filter.

Setting the Layout

The first step in configuring the Video Collage filter is to arrange the "holes" (referred to as *tiles*) using the controls in the Inspector.

1 Click the Effects Inspector to reveal the Video Collage controls.

2 Click the Preview Layout checkbox.

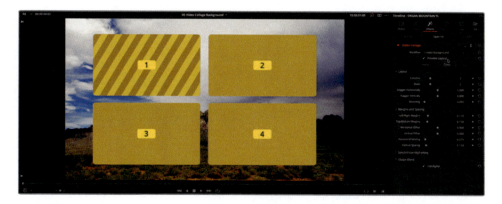

Each of the tiles is now clearly highlighted and denoted by a number. The actual number of tiles is dictated by the Columns and Rows controls in the Inspector. The tile with the shaded lines indicates the currently selected tile.

3 By default, the number of Columns is set to 2, and the number of Rows is also set to 2. This is the correct layout for this exercise.

4 In the Inspector, click the Tiles button to switch to the tiles controls.

These controls allow you to customize the tiles either as a group or individually. Since you won't be needing Tile 4, you might as well remove it from the layout.

5   In the Active Tile pop-up menu, choose Tile 4.

In the layout, Tile 4 becomes the active tile, as indicated by the shaded lines in the layout.

6   Click the Manual Tile Management checkbox.

7   Click Delete Tile to remove Tile 4 from the layout completely.

With no Tile 4, Tile 3 now becomes the currently active tile by default.

8   From the Active Tile pop-up menu, choose Tile 2 to make it the currently active tile.

9   Click the disclosure arrow for Custom Size/Shape to reveal the controls.

Currently, Tile 2 is occupying only one part of the grid: Column 2 and Row 1.

10   Change the End Row value to 2 to expand Tile 2 into the space formerly occupied by the deleted Tile 4.

With the layout of the tiles now set, you will customize their look all together.

11   Click the disclosure arrow next to Tile Styling to reveal the controls.

12   Increase the Tile Border to about 0.025 to add a consistent border around all the tiles.

TIP You can uncheck the Apply To All Tiles checkbox to further customize the border settings for the currently selected tile, if required.

13 Click the disclosure arrow for the Drop Shadow controls and increase the Strength value to around 0.50 and the Drop Angle to about 90.0, the Drop Distance to about 0.03, and the Blur to about 0.50 to customize the drop shadow around each tile.

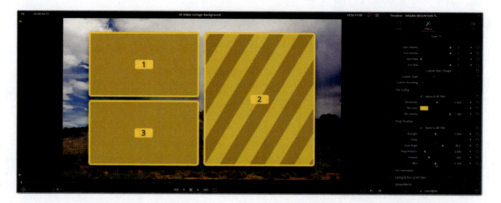

Finally, one of the more powerful aspects of the Video Collage filter is the ability to create animated intros and outros quickly and easily for each tile. These animations can be either manually keyframed or automatically generated over a customizable duration. There are four types of animations to choose from: Fade, Fly, Shrink, and Rotate. In Create Background mode, however, only the holes created by the tiles are animated, rather than their content, so Fade or Shrink are good choices here.

14 Click the disclosure arrow to reveal the Tile Animation controls, change the Animate pop-up to Intro Only, uncheck the Shrink checkbox, select the Fade checkbox, and change the Duration slider to 24 to create a 1-second fade for the tiles (this timeline is set to 24 fps).

15 Play the timeline to preview the effect so far, including the fade in and fade out uniformly applied to all tiles.

With the parameters of the Video Collage effect set, now it's time to fill the tiles.

Resizing the Content

Now that you have used the Preview Layout to set up the grid, look, and animation of the Video Collage effect, it's time to turn off the Preview mode and scale the underlying clips to fill the holes in the background clip.

1   Position your playhead about halfway through the timeline (at about 2:00) so that you can clearly see the three tiles, after their 24-frame fade in.

2   At the top of the Video Collage controls, uncheck Preview Layout.

Once again, the clip on Video 3 becomes visible through the holes created by each tile. You now need to scale the clip to fit the hole created by Tile 1.

3   Select the clip on V3 and choose View > Viewer Overlay > Transform or click the timeline viewer's onscreen Transform controls.

4   Use the onscreen controls to adjust the zoom and position of the clip on V3 so it fits in the hole created in the top left of the grid.

5 Select the clip on V2 and, again, use the onscreen controls to size and position the clip in the hole created by Tile 3, with the girl dominant in the frame.

6 As the clip's edges extend untidily beyond the edges of the tile, choose View > Viewer Overlay > Crop, or change the onscreen controls to Crop, and use the onscreen controls to remove the excess from around the outside of the tile.

7 Finally, choose View > Viewer Overlay > Transform, or change the onscreen controls back to Transform, select the clip on V1, and resize the interview in the hole created by Tile 2.

8 When you have each clip placed correctly within the correct holes, choose View > Viewer Overlay > Toggle On/Off, or press Shift-` (grave accent), to turn off the onscreen controls and review the timeline.

Awesome! However, as always, there's one last final touch to add.

Stacking and Reordering Effects

The effect you've built using the Video Collage filter is looking great, but to really emphasize the foreground clips you may want to add a slight amount of blur to the background.

1 Deselect all clips in the timeline and, in the Effects Library, scroll to the top of the Filters list to the Resolve FX Blur category. Double-click the Gaussian Blur filter to the clip on V4.

Ah…By adding the blur filter, you're simply blurring everything on that clip—including the edges of the tiles and the drop shadow!

You'll need to reorder the effects on the clip to keep those elements sharp and just blur the image of the mountains.

2   Click the Effects tab in the Inspector.

Notice that the Gaussian Blur filter is listed below the Video Collage filter because it was applied after the Video Collage.

3 Click the Up Arrow control for the Gaussian Blur filter to move it above the Video Collage filter.

The background image remains blurred, but now the edges of the tiles are not blurred because you have now specified that this is the order in which these effects should be applied.

4 Click the Gaussian Blur filter in the Effects Inspector to access its controls and adjust the amount of blur to your liking.

> **NOTE** In the edit page, only the controls for one filter can be displayed in the Effects tab at any one time; however, you can always access the controls for other effects by clicking their names as you have just done for the Gaussian Blur filter.

Excellent. You've seen how the Video Collage filter can be used to create effective picture-in-picture effects over and above just using the transform controls available in the Video Inspector in Create Background mode. Next, you'll explore how you can use it in Create Tile mode.

Creating Tiles with Video Collage

An alternative way of employing the Video Collage filter is to use it to create and animate clips as individual tiles, much like you did in the first exercise in this lesson. This follows much more the traditional way of laying out a composite by having the background layer on the lowest video track and subsequent layers edited above this.

As before, it's best to set up the Video Collage on one clip using the Layout Preview mode. Once you're happy with the layout and animations, you can simply copy the effects to other clips where you can then adjust them further.

Much like editing in general, there may seem to be many complex steps involved, but once you have things set up, the payoff is well worth it!

1   From the timeline viewer pop-up menu, select the **Video Collage Tiles** timeline.

This timeline has been set up with the same "background" clip on Video 1, and a logo that fades in on Video 2. Video 3 through 5 are currently disabled, but all have clips that you will use to build a custom animated intro for Organ Mountain Outfitters.

2   Enable track V3 in the timeline and review the shot of the girl in the hat.

3   Move your playhead to the point at which she has turned to look at the camera (at about 3:00).

4   From the Effects Library, locate the Resolve FX Transform group of filters and apply the Video Collage filter to the clip on V3.

As before, the filter defaults to showing a 2 x 2 grid displaying the clips on the lower two video tracks.

5 Select the clip on V3 and click the Effects tab in the Inspector.

6 Change the Workflow pop-up menu to Create Tile.

The girl in the hat now displays as a picture-in-picture.

7 Click the Preview Layout checkbox to view the familiar tile preview.

8 Change the number of Columns to 3 and number of Rows to 1.

9 Increase the Rounding to 1.0 to create three circular tiles.

10 Change the Vertical Offset to about 0.20 so the tiles sit above the mountains and increase the Horizontal Spacing to about 0.125 to reduce the size of the tiles.

> **NOTE** The size of the tiles is based on the Left/Right Margins and Top/Bottom Margins, with the outermost tiles placed at these margins and any additional tiles distributed evenly between those outermost tiles. Therefore, increasing the Horizontal Spacing between each tile results in each tile being made smaller, since the outer margins haven't changed.

11 Click the Tiles button and open the Tile Styling controls.

12 Increase the Tile Border to about 0.025 and then click the Tile Color chip and use the system Color Picker to select a white border to match the Organ Mountain Outfitters logo.

13 Open the Drop Shadow controls and change the Strength to about 0.350, the Drop Angle to 90.0, the Drop Distance to about 0.035, and the Blur to about 0.5 to create a subtle, diffused drop shadow.

14 Deselect Preview Layout to display just the current title (Tile 1) and the clip of the girl turning toward the camera.

15 Open the Resize Content controls and adjust the Pan to about 0.06, the Tilt to about 0.35, and the Zoom to about 0.85 to reframe the girl in the circle, not forgetting to keep the Organ Mountain Outfitter's clothing brand in shot!

The filter is now ready to be copied to the other clips in the tracks Video 4 and 5.

Copying and Pasting Attributes

The easiest way to apply the same effect, with the same settings, to the other clips is to use the copy and paste attributes commands, after which you only need to make one or two adjustments to the copied filters for the new clips.

1 Click the clip on V3 to ensure that it's currently selected, and then press Command-C (macOS) or Ctrl-C (Windows) or choose Edit > Copy to copy the clip.

2 Shift-click the Enable Video Track button for V4 or V5 to enable all disabled tracks.

3 Select the clips on V4 and V5 and press Option-V (macOS) or Alt-V (Windows) or choose Edit > Paste Attributes.

4 In the Paste Attributes dialog, select Plugins and click Apply.

The Video Collage filter is applied to the selected clips with exactly the same settings as the original clip it was copied from. You will need to make a few changes to these copied filters.

5   Command-click (macOS) or Ctrl-click (Windows) the clip on V5 to deselect it, leaving just the clip on V4 selected.

6   Click the Effects tab in the Inspector to reveal the Video Collage settings for the selected clip.

7   In the Manage Tiles controls, change the Active Tile pop-up menu to Tile 2.

The clip on Video 2 appears in the position specified for Tile 2.

8   Open the Resize Content control and change the Pan to about 0.07, the Tilt to 0.0, and the Zoom to 0.6.

9   Select the clip on V5, change the Active Tile pop-up to Tile 3 and, in the Resize Content controls, change the Pan to about 0.009, the Tilt to 0.0, and the Zoom to about 0.585.

Now that you have each of the tiles styled, it's time to add some keyframed animation.

Although the Video Collage effect has its own animation parameters, it's often easier to animate multiple elements together in the same compound clip. That way, if you wish to change the animation, you only have one set of keyframes to worry about.

10   In the media pool, select the Compound Clips bin.

11   Select the clips on V3, V4, and V5, and then choose Clip > New Compound Clip or right-click them and choose New Compound Clip.

12 Name the new compound clip **OMO Tiles** and click Create.

The three clips in the timeline are collapsed into a single compound clip, leaving two empty timeline tracks.

13 Right-click the timeline track headers and choose Delete Empty Tracks to remove the redundant video tracks.

NOTE The empty audio track will not be deleted. Each timeline must contain at least one video and one audio track.

14 Place the timeline playhead at the position just as the logo on V2 is almost fully faded in and, with the **OMO Tiles** compound clip still selected, click the Position keyframe in the Inspector.

15 Return the playhead to the start of the timeline and change the Y Position value to about 600, or until the tiles have moved fully out of the top of the timeline viewer, adding a new keyframe to this position automatically.

16 With the compound clip still selected in the timeline, Choose Clip > Show Curve Editor or press Shift-C.

17 Click the Curve pop-up menu and change the current curve to Position Y.

18 Select the second keyframe in the Curve Editor and click the Ease In button to add a Bézier handle to the left side of the keyframe.

19 Drag the Bézier handle down slightly to flatten the curve, slowing the animation as it approaches the keyframe position.

20 To change the timing of the animation, select the compound clip in the timeline and choose Clip > Show Keyframe Editor or press Shift-Command-C (macOS) or Shift-Ctrl-C (Windows).

21 Drag the second keyframe in the Keyframe Editor to the left for 15 frames or until the graphic on V2 has faded in completely.

22 Press Shift-C to close the Curve Editor and press Shift-Command-C (macOS) or Shift-Ctrl-C (Windows) to close the Keyframe Editor and play back the timeline to review the animation.

> **TIP** If you wish to manually keyframe the individual tiles created using the Video Collage effect, you'll need to nest each of the clips in their own compound clip in order to adjust the render order of the Video Collage and Transform controls

Well done! You have successfully completed this lesson and can now use some of the effects included with DaVinci Resolve 18 to complete common tasks required of editors on a daily basis.

> **NOTE** A series of finished timelines for each of the exercises in this lesson are available for you to import into this project from R18 Editors Guide/Lesson 07/ Timelines.

Lesson Review

1   Which DaVinci Resolve FX can be used to key blue- or green-screen footage over a background?

    a)   Luma Keyer

    b)   HSL Keyer

    c)   3D Keyer

2   True or false? The only way to determine whether a clip has a Resolve FX applied is to open the Inspector.

3   Which Resolve FX can be used to easily adjust a clip's pitch, yaw, width, and height values using intuitive onscreen controls?

    a)   Distort

    b)   Perspective

    c)   Transform

4   Which Resolve FX can be used to quickly create complex picture-in-picture effects?

    a)   Grid

    b)   Video Collage

    c)   Blanking Fill

5   True or false? Render in Place creates a video file using the render cache options in Project Settings.

Answers

1 b) and c). The Luma Keyer has no controls for selecting hue and saturation values.

2 False. The timeline clip displays a small FX badge next to the clip name.

3 c) Transform.

4 b) Video Collage.

5 False. Render cache and Render in Place use different settings.

Editing and Mixing Audio

Your project's soundtrack is an essential part of the overall audience experience. You could have the most wonderful editing, fantastic effects, and superb grading on your film, but if your audience can't hear what's happening clearly, they won't be able to engage with your story and will quickly switch off. And that's as true today for films, TV shows, and online social media videos as it's ever been.

In this lesson, you'll explore some specific techniques for audio editing, sound design, and final mixing for your timelines.

Time

This lesson takes approximately 60 minutes to complete.

Goals

Preparing the Project

For this lesson, you'll work on the soundtrack for the short scene from Sync that you previously edited in Lesson 4, "Editing a Dramatic Scene."

> **TIP** Ideally, you'll want to have a good set of speakers or headphones connected to your computer for this lesson to appreciate the audible subtleties.

1   In the Project Manager, right-click and choose Import Project. Navigate to R18 Editors Guide/Lesson 08, select the **SYNC SCENE AUDIO.drp** project, and choose Open to import the project into your Project Manager.

2   Open the project and relink the media files.

3   Choose Workspace > Reset UI Layout.

4   From the Select Timeline pop-up menu, select the **SYNC SCENE FINAL MIX** timeline. Adjust the timeline track heights if appropriate so you can see as many of the timeline clips as possible.

5   Play through the **SYNC SCENE FINAL MIX** timeline to reacquaint yourself with the scene.

Wow! There's a lot going on in this timeline! You should be familiar with this scene between Doctor Kominsky and Agent Jenkins, which you cut in an earlier lesson. This is a slightly refined version of the scene with several lines of dialogue having been removed to help the scene's pacing. Some additional elements have also been added to enhance the soundtrack; the audio clips on A4, for example, are a radio call being received by the FBI agents about the developing situation, and there are also music and sound effects on A6–A8 that drown out Agent Jenkins's final lines.

To get a better sense of how the audio is currently working, you'll mute some of the tracks.

6   Open the Index window and select the Tracks tab.

The Tracks Index gives you a more compact view of the track controls, including the track names.

7   Click the Mute button for tracks A4, A7, A8, and A9, and then play the timeline again to preview the scene.

Notice how the timeline has been arranged with each character in the scene having their own audio track; even the assistant who has only one line in this scene (and was recorded on the Doctor's audio track) has a track dedicated to that line. This is to aid the audio mixing process, which you will get to in good time.

When Should You Start Mixing?

Up to now, you've learned how to use DaVinci Resolve to edit audio at the subframe level, adjust audio levels, add fades, etc. However, most audio work occurs toward the end of the editing process, and that's why this lesson is the penultimate chapter of this book.

While you will no doubt make basic adjustments to audio levels throughout the editing process, you don't really want to spend too long perfecting the audio mix until the editing has been completed. This is the point that's usually referred to as *picture lock*, where the director or client is happy with the work so far and the fine-tuning can begin. If you start mixing the audio too early, you may find yourself in a position where you'll have to cut out a part of the scene, or maybe the entire scene completely, in which case all your hard work and the time taken to achieve it will be wasted. This is also the reason why *grading* the picture also occurs once the scene is picture locked and no further changes should be made.

In practice though, while picture lock may be aspired to, many edits may need further tweaking after a mix or grade has been started (or, in extreme cases, completed). In these cases, the benefit of having the editing, mixing, and grading tools together inside the same application becomes obvious, as Resolve allows you to move seamlessly from one process to another and back again just by clicking the appropriate page.

The vast majority of the timelines you will work on will have three main audio components: *dialogue* (words spoken by actors, interviewees, or in voiceover/commentary), *effects* (sounds that occur onscreen or offscreen, such as a door slamming, the tapping of an iPhone's keyboard, an aircraft approaching from a distance, or wolves howling in the wilderness), and *music* (which sets the tone of the scene).

Quite often, there might be several clips across different audio tracks for the same type of audio. For example, this timeline has four tracks for dialogue clips: one for each of the speaking actors and a fourth for the radio operator's lines. There are four effects tracks: one for the audience applause at the start of the scene, one for the sounds of the crowd in the room, and then two more for the crash and bang effects at the end of the scene. There's also one music track for the mixed score.

The audio in each of these groups can be *internal* to the scene or *external*. Internal audio (often referred to as *diegetic*) comes from a source the viewer can place somewhere within the scene, such as a specific speaker or music playing from a radio. It doesn't always need to be emanating from a visible source to be classed as diegetic sound; the sounds of cars, buses, sirens, etc., can indicate that the scene is taking place at a busy city intersection, even though not a single vehicle may be seen onscreen! External audio (called *non-diegetic*) is anything in the soundtrack that couldn't possibly come from anything within the onscreen world, such as a narrator's voiceover or music used to enhance the audience's emotion.

Again, this scene contains a mixture of *diegetic* and *non-diegetic* audio. Can you identify which is which?

The applause as the doctor walks off the stage in the scene's first shot is obviously meant to be made by a large crowd of people she's just been addressing, while the sounds of people speaking in the background are made by the extras surrounding the doctor and the agents throughout the rest of the scene. (This *ATMOS* track you added in Lesson 4—the nondescript audio of a crowd—also has the advantage of providing a consistent background noise, or *room tone*, that helps fill the gaps in the soundtrack left during the dialogue clips.) However, the sounds of impacts and the music toward the end of the scene are obviously meant only for the audiences' ears, to provide an emotional resonance to the end of the scene.

Replacing Lines from Other Takes

A common problem that editors often face is that a take might be great visually, but the audio might be less than optimal. Indeed, many takes are often chosen initially on the basis of their visuals rather than their audio. In these cases, it's often easier to replace part of the audio in the timeline with the audio from another take.

1   Select the Markers tab in the Index and double-click the second marker named **REMOVE CLAP?**.

The playhead jumps to the second red marker in the timeline.

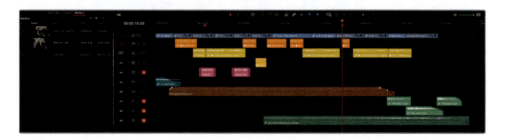

> **TIP** Pressing Shift-Up Arrow or Shift-Down Arrow jumps the playhead to the previous or next marker, respectively.

2 Press / (slash) to play around the current playhead position, paying attention to the doctor's line.

Ouch! It sounds as though Doctor K hits herself as she brings her hand down! It's very distracting but probably very difficult to edit out (which should have been your first instinct). Instead, you will use Resolve's audio track layers to replace the line with another take.

3 Zoom in on the current playhead position and use Shift-Mouse scroll to increase the height of the audio tracks.

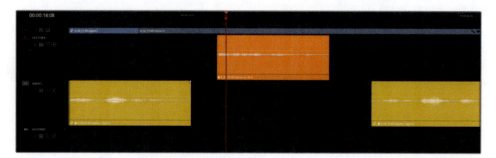

4 Choose View > Show Audio Track Layers.

All the timeline tracks appear to shrink in height. But look closer, and you will see a space appear above the clips for each track, separated by a thin dividing line. These are the audio track layers.

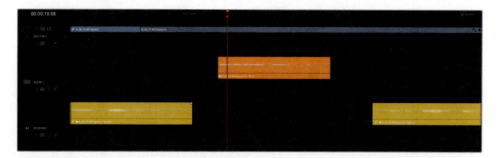

NOTE Audio track layers function very similar to the way you work with video tracks. You can add any amount of audio clips to the audio track's layers, and they will appear one above the other. However, only the topmost layer is the clip that plays. The topmost clip always has priority; you can't "mix it" with a clip on a lower level, although you can use audio fades on the uppermost clips to fade them in from or out to a lower clip.

5 In the media pool, select the Video bin in the Dialogue Clips bin and double-click the clip **A_S8_T2 WS Agent J** to open it in the source viewer.

This clip has a blue duration marker locating the alternative take of the line of dialogue.

6 Choose Mark > Set In and Out from Duration Marker to add the required In and Out points.

7 Press Option-/ (slash) on macOS or Alt-/ (slash) on Windows to play from the In to Out points.

This is a clean version of the same line of dialogue.

8 In the timeline, deselect the track source controls for A2 or press Option-Command-2 (macOS) or Alt-Ctrl-2 (Windows).

9 Disable Snapping and use the source viewer Overlay button to drag the audio clip into A1 above the existing clip (the doctor's lines are on the first source channel).

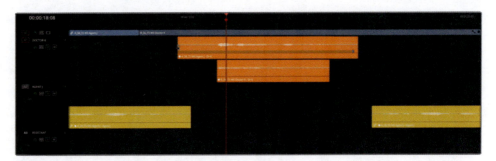

10 Trim the new clip so it starts at the waveform where the doctor begins speaking and align it with the beginning of the original clip on the track layer below.

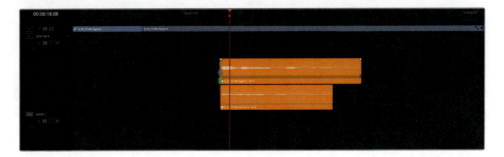

11 Press / (slash) to preview the new line of dialogue.

The clap has vanished!

12 Click and drag the original dialogue clip up and above the new clip to swap their order in the track layers.

TIP Hold Shift when changing the layer order of the clips to prevent either moving horizontally in the timeline.

13 Press / (slash) again to preview the original take.

14 Change the order again so the new take sits above the old take on the track layers.

You'll probably get away with using the audio from this different take, as the actress playing the doctor delivered her lines at roughly the same speed and it's doubtful anyone would ever spot a discrepancy between her lip movements and the words spoken. However, if you look closely, the waveforms of the uppermost layer don't fully align with the original take's audio. But you can easily fix this by adjusting the speed of the audio clip.

15 Right-click the topmost audio clip and choose Change Clip Speed.

16 Change the speed to 115%. Ensure that Ripple Timeline is not selected but Pitch Correction is selected and click OK.

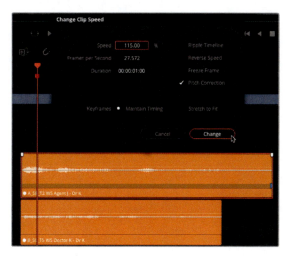

The speed of the audio clip is increased, and the audio pitch is automatically adjusted so the doctor's new audio is now in near-perfect sync, and she doesn't sound like she's inhaled helium.

17 Choose Timeline > Flatten Audio Track Layers to remove the unwanted take from the timeline.

18 Trim any excess audio from the newly added clip, and then choose View > Show Audio Track Layers to hide the track layers.

19 Select the Full Extent Zoom and adjust the audio track heights so you can see all the audio tracks in the timeline again.

Audio track layers are a useful feature of working with audio in DaVinci Resolve timelines. They can be used for many different things, such as quickly switching between different voiceover takes or auditioning different music cues for a scene. Used this way, they are very similar to using the Take Selector, which cannot be enabled for audio-only clips.

Balancing the Dialogue Clips

Of the three general types of audio, probably the most important one is dialogue: if your audience cannot hear what's being said, they will not be able to follow the story, whether that is the story told in a drama scene such as in this lesson or through an interview, as with the *Organ Mountain Outfitters* footage you edited in previous lessons. Think about how much information is communicated in what the people onscreen are saying. The only time when dialogue is not the most important element of the soundtrack is for a video that is cut entirely to music (such as the Jitterbug Riot footage in Lesson 5, "Multicamera Editing") or when the dialogue is purposefully meant to be inaudible.

With that in mind, it's not surprising to learn that, as when starting to edit a piece together, dialogue is usually the best place to start when it comes to the audio mixing process.

1   Play the timeline again, and this time listen carefully to the dialogue clips.

The audio for this scene has been recorded quite well, and the audio editing means just the parts of the clips with spoken words are included. However, you need to set all the dialogue clips to the same general level.

2   At the top right of the interface, click the Mixer button.

The edit page audio mixer opens to the right of the timeline. If you don't see the mixer at full width, drag the left edge of the mixer to the left in order to reveal all the controls.

> **TIP** If you need more space along the bottom of the interface for the timeline and the mixer, you can use the Shrink button to the left of the Media Pool button in the interface to close the Index and use a half-height media pool.

The mixer contains a strip of controls for each audio track in your timeline, and an additional strip for the master control for the entire timeline, currently called Bus 1. In DaVinci Resolve, each audio clip can be adjusted within the track, as you've been doing so far. Each track can then be further adjusted by using the mixer before it is passed to the output bus. It is the output from Bus 1 that you hear playing through your speakers or headphones. Any adjustments made to the track controls in the mixer affect any clips on that track.

3   Play the timeline once more, and this time keep an eye on the audio levels for each track as the clips play.

You should see that the levels for Doctor K's audio clips are a lot lower than those of Agent J.

At this point, you might be tempted to reach for the audio level control for the track, but you first need to adjust the clip levels as much as you can; just because one clip is at one level doesn't mean that they all are!

4   Play the first dialogue clip, which is on track A1, and look at the levels displayed in the track's control strip in the mixer.

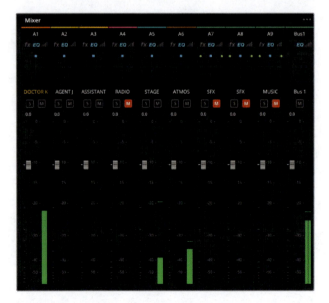

The clip's levels barely peak above -20 dBFS! This clip is a good candidate for increased audio levels.

5 Increase the volume of the clip in the timeline until the tooltip reads about +8 dB and play the clip once more.

> **TIP** Hold Shift when adjusting the volume of the clip for more precision.

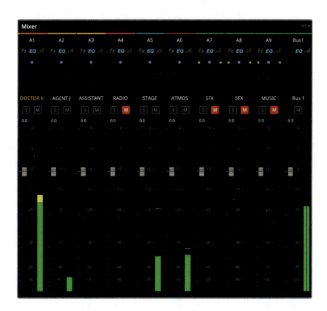

The volume of the clip has increased and now sits in the yellow region of the track level, peaking around -15 dBFS and sounds much better compared to the following clip of Agent J.

How Much Should You Adjust Your Clip's Audio Levels?

The mixer in DaVinci Resolve also gives you a good indication as to where your audio levels should be set, with each meter showing a range of levels color-coded in green, yellow, or red. For your current purposes, it really doesn't matter what level you choose *as long as you are consistent*—that is, you set the volume for the clips on the same track to the same level.

A good guide for dialogue levels is to set them within the yellow area, which is between -18 dBFS and -8 dBFS on the audio mixer. Quiet, softly spoken lines might be toward the bottom end of this range, and shouted lines will be toward the higher end, with normal dialogue levels being set around -12 dBFS.

Normalizing Audio Levels

At this point, you now need to balance all the dialogue clips so they play at a similar level. Remember, your audience must clearly hear each line of dialogue. You can, however, employ normalization across the dialogue clips to help with this process. It's not a "Get Out of Jail Free" card in that it doesn't necessarily *fix* all your audio levels, but it can help streamline the process on many occasions.

1   Select all the clips on the first three audio tracks, A1, A2, and A3.

2   Right-click any of the selected clips and choose Normalize Audio Levels.

    The Normalize Audio Levels window appears.

3   Leave the Normalization Mode set to Sample Peak Program, leave the Target Level set to -9 dBFS but change the Set Level option to Independent. Click Normalize.

The selected audio clips have their volume adjusted as indicated by the change in the waveforms. Notice how the waveforms for the different clips are now all similar sizes, indicating that they are roughly the same level.

4   Once more, play the timeline and see the change in the audio levels in the mixer.

Now all the levels seem to be playing at a much more consistent level. However, as mentioned earlier, the process of normalizing the audio levels isn't a fix in itself; it just gets you in the ballpark. You now need to go through and ensure that the clips' levels are where you want them.

5   Play the third clip on A1. This is where the doctor mutters to her assistant.

The clip has been raised to the level you specified when you normalized the clips, but it's totally unsuitable for this particular clip.

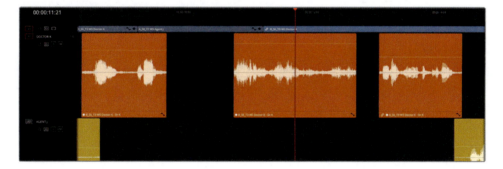

6 Lower the clip's volume curve to about 12 dB, which puts the clip at the lower values of the yellow target area on the meters.

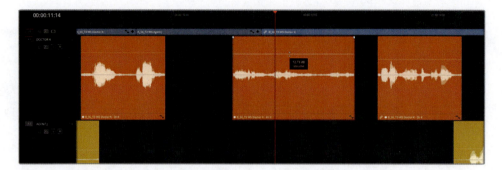

You can always refine the levels further by adding keyframes.

7 As the clip is still a little noisy at the end, add a couple of keyframes after the doctor has delivered her line by Option-clicking (macOS) or Alt-clicking (Windows) the volume curve and lowering the portion after the second keyframe to about 4 dB to push the rustling noises back into the overall mix.

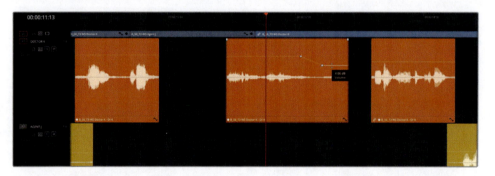

Continue to balance each of the dialogue clips across the first three tracks so they are consistent throughout the scene.

Introducing the Dialogue Leveler

DaVinci Resolve 18.1 introduced a new audio track control to help smooth out dialogue audio levels that may be a bit too low or too high. Like Normalization, the Dialogue Leveler is not a fix in itself but can help with the often tedious need to adjust multiple clips' levels using numerous keyframes.

Ideally used as part of the Fairlight page's more comprehensive mixing environment, where it is available as a Fairlight FX in the mixer, the Dialogue Leveler is also available in the edit page by opening the Inspector and clicking the audio track header in the timeline.

The Dialogue Leveler has some self-explanatory controls for lifting soft portions and reducing loud portions of the dialogue, together with a slider to adjust overall output gain.

Again, like the Normalization process, the Dialogue Leveler is not a solution for wrong audio levels. Instead, think of it as a way of bringing your dialogue levels into line.

NOTE If you need to catch up before moving to the next step, select the Timelines bin and choose File > Import > Timeline, navigate to R18 Editors Guide/Lesson 08/Timelines and select SYNC SCENE FINAL MIX CATCHUP 1.drt and click Open.

Enhancing the Scene

Now that the dialogue has been repaired and balanced, it's time to turn your attention to the sound design for this scene. In the script, this scene is set during a break at a busy technology conference. While the **Atmosphere wild sound** clip on A6 offers some sense of the other people in this room, it's supposed to be in a room full of people chatting, and there's a distinct lack of *atmosphere*, which you will now inject. To begin with, you'll duplicate the **Atmosphere wild sound** clip to increase the number of people in the room.

1   Right-click on the timeline header controls for A6 and choose Add Track > Mono.

    A new A7 audio track is added to the timeline, below the A6.

> **NOTE** You can always reposition a track's position in the timeline by right-clicking the track header and choosing Move Track Up or Move Track Down, as appropriate. Alternatively, you can use the Tracks Index panel to freely drag audio tracks to a different order.

2   In the mixer, double-click the new Audio 7 track name and type **ATMOS** to rename the track.

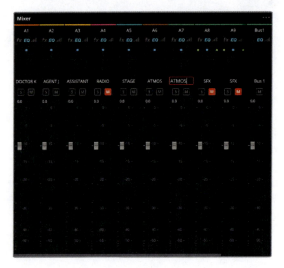

3   In the timeline, right-click the track header for A7 and choose Change Track Color > Chocolate.

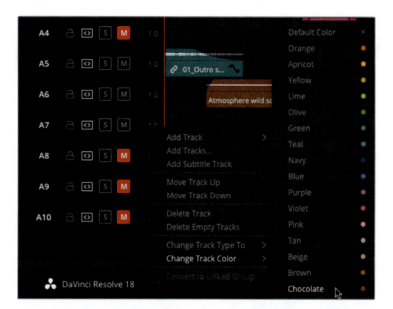

4   Press and hold the Option key (macOS) or the Alt key (Windows), select the **Atmosphere wild sound** clip in A6, and drag a copy of the clip into A7.

> **TIP** If you don't have snapping enabled before you start duplicating the clip, press and hold Shift together with the above step before you release the mouse to ensure that you constrain the horizontal position of the clip in the timeline.

Listen to the change this makes. Maybe you'll agree that it doesn't sound as though you've increased the amount of people in the crowd—rather, they are just louder.

5 Press T to activate Trim Edit mode and slip one of the two Atmosphere wild sound clips by about -10 seconds.

6 Finally, right-click the Atmosphere wild sound clip in A7 and choose Change Clip Speed. Change the Speed to 90%, uncheck the Pitch Correction option, and click Change, before reviewing the changes.

You're getting there. By offsetting the timing and adjusting the speed and pitch of one of the clips, it sounds as though there are so many more people in the room now, even though only a handful of people are actually onscreen. Congratulations! You have increased the crowd size without having to pay for additional actors. The producer will be pleased.

Panning Tracks in Acoustic Space

Pan controls enable you to choose where a track's audio is placed within a panoramic sound field. They enable you to adjust the spatial arrangement of audio elements just as a cinematographer composes the visuals of a shot. Mono tracks can be precisely located to sound as if they come from an offscreen source or from somewhere within the frame. DaVinci Resolve includes advanced pan controls in the mixer that support both 2D (stereo) and 3D (surround) audio placement.

In this exercise, you'll use the pan controls in the mixer to "widen" the audio from the atmos tracks so they don't "crowd" the dialogue tracks.

1   Return your playhead to the start of the scene and play again, listening carefully with your eyes closed. Can you pinpoint where the dialogue and atmos audio is all coming from?

Your eyes may tell you where things are onscreen, but when you close them everything seems to be coming from directly in front of you. Of course, the crowd should be all around you, with only the dialogue from the actors coming from directly ahead.

Because the audio is playing from a series of mono tracks, the sound from all these tracks is playing equally from both output channels in the timeline, and therefore equally from both of your speakers. This makes them sound centered within the sound environment.

2   Continue playing the scene and, using the mixer's Pan controls, drag the blue handle for the A6 track to the upper-left corner of the panning control in the control strip.

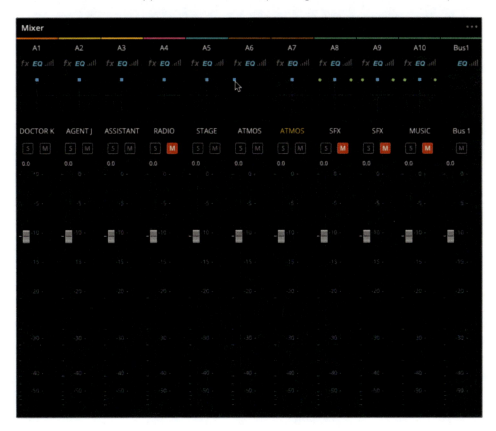

Can you hear the change? It sounds like you're pushing part of the crowd to the edge of the room.

3   Drag the A7 track's blue panning handle to the upper-right corner of the control.

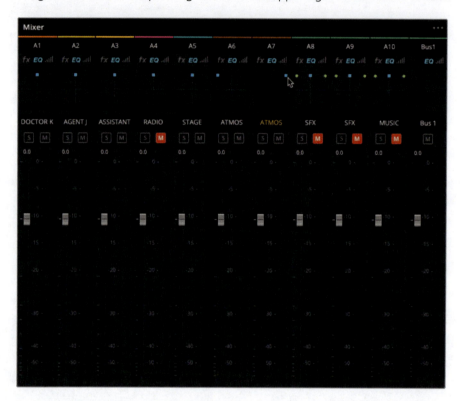

4   Play the timeline from the beginning to hear the atmos tracks panned to the far left and right of the acoustic space.

In just a matter of seconds, you filled the far reaches of the acoustic space with the atmos, making the crowd seem to spread out within a much larger room. Crucially, you've also moved the sound of the crowd away from the actors' audio tracks, thereby making it easier to hear the all-important dialogue.

> **TIP** For more control over a track's pan, double-click the Pan control in the mixer.

5 In the mixer, change the name of the A6 and A7 tracks to **ATMOS-L** and **ATMOS-R**, respectively.

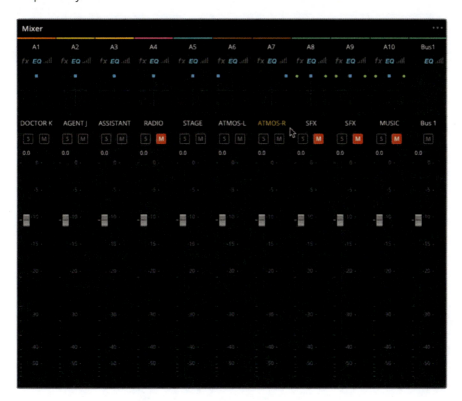

When you're finished panning tracks to compose the acoustic space within your scene, you can move on to finessing the levels of those tracks. In audio post-production, volume control is an ongoing process right up until you output the final mix.

Increasing the Audience

To practice the above steps, repeat the process on the opening applause clip:

1 Create a mono track below the existing STAGE track. Rename this track **STAGE-R** and color code it Teal to match the A5 track.

2 Rename the STAGE track **STAGE-L**.

3 Duplicate the audio of the 01_Outro shot from previous scene clip to this new track.

4 Slip the applause audio of one of the clips and adjust the track pan so that the applause fills the room.

5 Try adjusting the speed of one of the applause clips, with or without pitch correction.

Adding More Atmosphere

The script for this scene calls for the FBI agents to interrupt the doctor at a high-brow conference with her colleagues. Now that you have increased the number of people in attendance, it's time to add a little more nuance to the scene.

1   Select the Foley bin in the media pool.

This bin contains the original **Atmosphere wild sound** clip, together with another audio clip, **Piano dry.aif**.

2 Double-click the **Piano dry.aif** clip to load it in the source viewer and play the clip to listen to it.

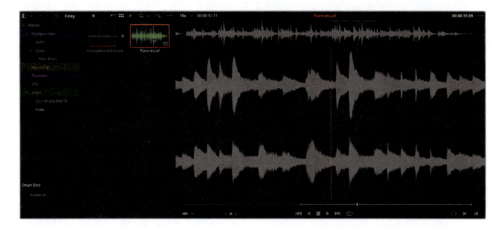

This clip is a stereo recording of a piano being played in a studio. A portion of the clip has already been marked with In and Out points. You will edit this clip into a new track to add to the overall ambience of the scene, as if there were a pianist somewhere offscreen tinkling the ivories for the delight of the conference attendees.

3 In the timeline, right-click the track header for A8, the ATMOS-R track, and choose Add Track > Stereo.

A new, stereo audio track is added to the timeline and the mixer, labeled A9.

> **NOTE** Because you're adding a stereo clip to this timeline, it makes sense that it should be added to a stereo track to preserve the stereo effect in the mix. If you add a stereo clip to a mono track, only the first of the two stereo channels are used, and if you add a mono clip to a stereo track, the mono audio will only be played out of the first channel as you experienced briefly in Lesson 3, "Fine Cutting an Interview." Incorrectly created track types can be adjusted by right-clicking the track header and choosing Change Track Type To, and incorrect audio configurations for clips can be adjusted in Clip Attributes.

4   In the mixer, change the name of the new A9 track to **PIANO**.

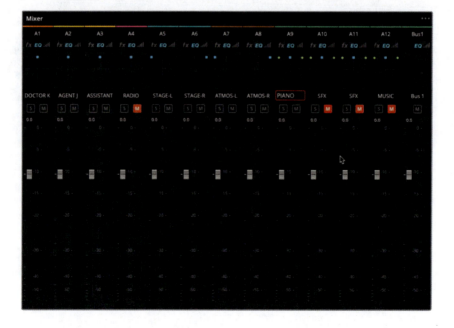

5 In the timeline, change the A9 track color to Tan and click the Track Destination Control to target the new track.

> **NOTE** Default keyboard shortcuts exist to quickly target any of the first eight video or audio tracks in the timeline (viewable by choosing Timeline > Track Destination Selection). For any track numbers above this, it's often easier to target them by clicking with the mouse.

You'll now use the duration of the clip on either of the ATMOS tracks to quickly mark a duration for your piano audio.

6 Select the clip on the A8 track and choose Mark > Mark Selection or press Shift-A.

In and Out points are added to the timeline for the duration of the selected clip.

7 Make an Overwrite edit.

8 Increase the size of the A9 track so you can see the waveform clearly and lower the Piano clip's volume by 25 dB to bring its level down to around the same level as the atmos audio (using the levels displayed for A9 in the mixer as a guide).

9 Using the Fade handles, add a short ten-frame fade at the start of the Piano clip and a 2-second fade at the end, changing the curve of the fade-out to create a subtle logarithmic fade for the piano.

Perfect. Now your conference crowd has some entertainment to listen to while the doctor and the FBI agents discuss the worsening situation.

Filling the Room

The piano has added some much-needed ambiance to the scene, but it still feels a little *dry*, since it was recorded in a sound studio with lots of foam cones on the wall that reduced any unwanted *sound reflections* from the studio walls. This is ideal for the foley clips because you can then add your own *reverb* to the audio using the built-in Fairlight FX to suit the scene.

1   Click the Effects Library and select Fairlight FX from the Audio FX category and locate the Reverb effect.

2   Drag the Reverb effect to the A9 track header in the timeline.

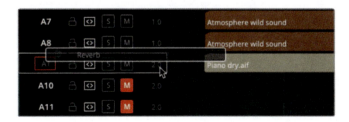

The Reverb FX controls window opens automatically.

3   From the presets pop-up menu in the top-left of the controls window, choose the Concert Hall preset.

4   Solo track A9 and play back the scene to hear the newly created sound reflections of the piano bouncing around the room.

> **TIP** Use the red active switch in the top-left of the Reverb controls to disable and enable the effect to hear the audio clip with and without the effect applied.

5   Close the Reverb effect controls window and un-solo the A9 track.

> **NOTE** In the edit page, to access the controls for an audio effect applied to a track, click the track header in the timeline and open the Effects tab in the Inspector.

Adding an effect to a track rather than individual clips is an easy way to make sure that any clips on any part of the track will have the reverb applied. You can tell if a track has an effect applied to it by the FX badge in the track header in the timeline and at the top of the track's controls in the mixer.

Creating a Radio Effect

Now that you have some idea of the sort of work that goes into a scene's *sound design*, it's time to turn your attention to adjusting the audio that is supposedly coming from the agents' radios. In this case, you'll apply some adjustments to the clips' EQ before changing the track's level for the mix.

1   Open the Markers Index and double-click the thumbnail for the first marker in the timeline.

2   Adjust the timeline zoom so that you can clearly see at least the first three clips on the A4 track.

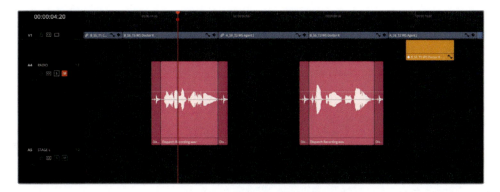

3   In either the timeline track header or the Tracks Index, click the A4 Mute button to unmute the track and click the A4 Solo button to solo the same track.

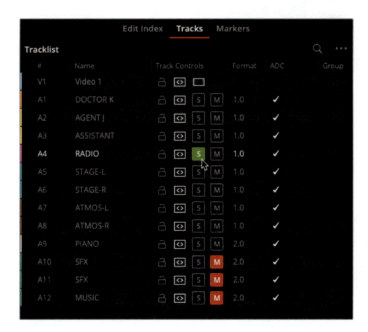

4 Press / (slash) to play around the marker.

This part of the timeline has a sound effect at the beginning and end of the line from the radio operator to indicate the radios receiving a signal. The actual lines from the radio operator were spoken by an actor and recorded directly into Resolve. However, as you did with the piano clip, you'll need to manipulate the audio to make it sound as though you're hearing this over a radio rather than through a studio microphone.

> **NOTE** See *The Fairlight Audio Guide to DaVinci Resolve 18* for more information on recording audio using the Fairlight page.

5 In the timeline, select the **Dispatch Recording.wav** clip on A4.

6 Open the Inspector, click the Audio tab, and scroll down until you see the Equalizer controls.

In DaVinci Resolve, each audio clip has a set of controls that includes a four-band equalizer that can be used for boosting or attenuating (reducing) different ranges of frequencies of the clip.

7   Enable the Equalizer for the selected clip.

8   In the Audio Inspector, move the Band 1 control right to about 650 Hz to cut the lower frequencies of this clip.

9   Move the Band 4 control to the left to around 1.5 kHz to cut the higher-end frequencies of the voice.

These changes limit the range of frequencies in the voice to just the midrange, making the clip sound a little tinny and artificial.

10 Drag the Band 3 control to the area between the Band 1 and 4 controls (around 1 kHz) and then drag it upward by around +10 dB.

This boosts the 1 kHz frequencies by 10 dB.

11 Play back the clip to review the changes you've made.

You can hear how the voice actor's dialogue has lost the warmth of the lower frequencies and the brightness of the upper frequencies, making it seem like it's emanating from a radio's tinny loudspeaker.

Copying and Pasting EQ

Having adjusted the frequencies of this first line of radio dialogue, you now need to apply the same adjustments to the second line. Just as you've done in previous exercises with video clips, you can copy and paste attributes between audio clips.

1   Select the **Dispatch Recording.wav** to which you've applied the EQ and choose Edit > Copy or press Command-C (macOS) or Ctrl-C (Windows).

2   Select the second line of dialogue in A4 (the 5th clip in the track) and choose Edit > Paste Attributes or press Option-V (macOS) or Alt-V (Windows).

3   In the Paste Attributes window, select the Equalizer settings and click Apply.

The EQ settings from the first line of dialogue are applied to the second line of dialogue.

4   Place your timeline playhead over the second line of radio dialogue and press / (slash) to hear the change.

Setting Track Levels

The radio dialogue is coming along nicely, but now you need to listen to it alongside the other dialogue tracks in this scene. In this case, the radio dialogue doesn't need to be heard clearly; it's more to add a sense of urgency to the FBI agents' enquiries.

1   Return your timeline playhead to the first red marker.

2   Click the Solo button for A4 to un-solo the track.

3   Press / (slash) to preview the line of radio dialogue as part of the overall sound design.

The radio dialogue is in the mix, but it's competing with the actual dialogue you need the audience to focus on, so you need to attenuate (lower) its levels so that it's less intrusive.

However, the dialogue clips are actually just two elements of the radio track, with the other being the clicking sounds of the radio connecting and disconnecting. If you were to adjust the level of the dialogue clips themselves, you'd also need to lower each of the clicking sounds too. This is why it's now easier to adjust the level for all of track A4.

4   In the mixer, use the Level slider to lower the audio levels for A4 by about -8 dB.

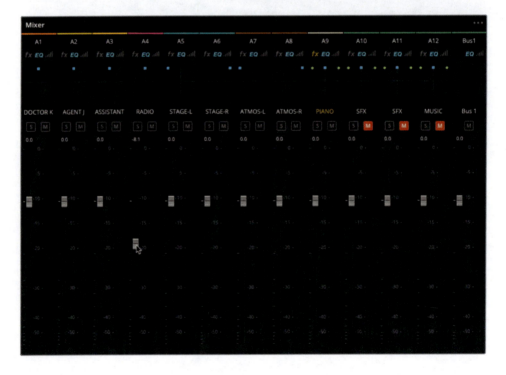

TIP You can see the relative adjustments you're making to the track's levels at the top of the slider controls.

5   Once you've made your change, press / (slash) once more to hear the results.

The radio dialogue is now better integrated into the scene without you having to adjust each of the clips' levels individually. This also demonstrates the advantages of having tracks dedicated to specific types of audio and how you can then use the mixer's track levels to further refine the levels for any track once the individual clips have been correctly balanced.

NOTE If you need to catch up before moving to the next step, select the Timelines bin and choose File > Import > Timeline, navigate to R18 Editors Guide/Lesson 08/Timelines and select SYNC SCENE FINAL MIX CATCHUP 3.drt and click Open.

Simplifying the Mix

Now it's time to think about how each track needs to be adjusted in the mixer. Because you already know that each of the dialogue clips is at the right levels, you shouldn't need to do a lot to those tracks. However, you might need to make further adjustments to each of the effects tracks—specifically, the ATMOS and PIANO tracks: A7, A8, and A9. While this is still a relatively simple scene with a limited number of tracks, it can still be advantageous to simplify this process by creating *additional buses*.

Buses provide a way in which you can add a further level of control to your audio mixing by routing different tracks through a *separate control strip* in the mixer, allowing you to adjust the output levels of those routed tracks with one set of controls.

To explore this process, you will create three additional buses for this scene's dialogue, diegetic effects and non-diegetic music elements.

You can create up to a total of 256 stereo buses (or 512 mono buses), which can then be used to control the audio from multiple tracks, vastly simplifying the process of mixing very complex timelines.

1   Choose Fairlight > Bus Format.

The Bus Format window opens, displaying the default bus for the timeline. This bus is used for the audio output of the entire timeline and is set to stereo by default.

Renaming the buses so you know what they are being used for is good practice.

2   Click the Name field for Bus 1 and change its name to **STEREO MAIN**.

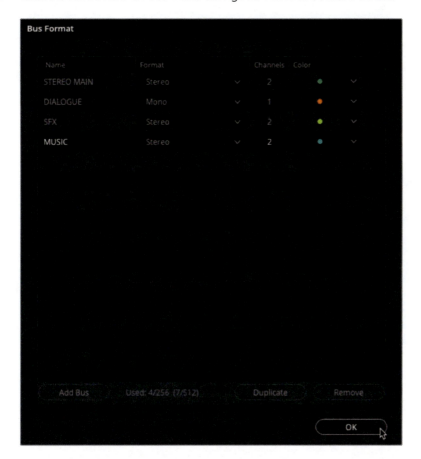

NOTE Even though you're on the edit page, you can still access some Fairlight features directly through the Fairlight menu. However, many options will be unavailable unless you're on the Fairlight page itself.

3   Click Add Bus to add a new bus call Bus 2 to this timeline.

You'll use this new bus for adjusting all the dialogue tracks in the timeline.

4   Change the name of Bus 2 to **DIALOGUE**.

> **NOTE** Short names for tracks are often useful since they're easier to read in the mixer or the track headers.

5   Leave the Format set to mono (this is for dialogue, after all) but change the color to Orange.

> **NOTE** Because a mixer bus doesn't contain any audio clips, changing its color really only has the benefit of identifying it in the mixer or the Bus Assign window (as you'll see later).

You now need to add two additional buses for the diegetic effects, and the non-diegetic effects and music.

6   Click the Add Bus button twice more to add a Bus 3 and Bus 4.

7   Change the name of Bus 3 to **SFX** and change the name of Bus 4 to **MUSIC**.

8   Change the format of both buses to Stereo.

9 Change the color of the SFX bus to Lime and the color of the Music bus to Teal.

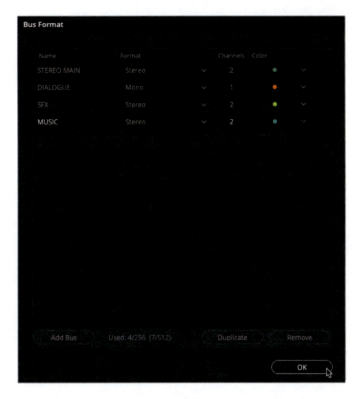

10 Click OK to save the changes and close the Bus Format window.

11 In the mixer, drag the dividing area next to the Bus 1 channel strip to reveal the controls for each of the buses.

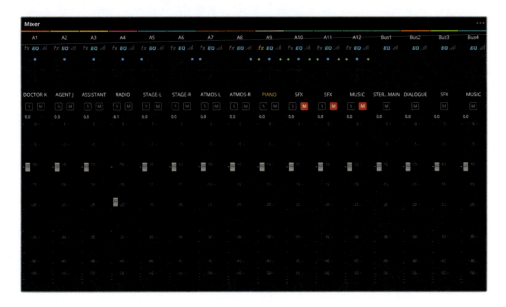

Assigning Tracks to Buses

You have successfully created three additional buses for this timeline. However, now you need to route the audio for each of the tracks to the appropriate bus, and then from each bus onward to the Stereo Main bus.

1   Choose Fairlight > Bus Assign to open the Bus Assign window.

The Bus Assign window is used to view and change the destination of the audio once it leaves a track. By default, all audio tracks are automatically routed directly to Bus 1, abbreviated as B1o (Bus 1 out). This is ultimately what you are listening to when the timeline plays.

2   In the Busses area, select the B1:STEREO MAIN Out and click Un-Assign All.

Now the tracks are no longer assigned to any destination, so you won't be able to hear them because they're no longer being sent to the B1 output bus.

3   In the Busses section at the top of the Bus Assign window, click the B2:DIALOGUE Out bus to select it.

4   In the Available Tracks area, click each of the dialogue tracks in your timeline: Doctor K, Agent J, Assistant, and Radio.

B2o (Bus 2 out) appears below the name of each track to identify which bus it is being routed to.

> **NOTE** Tracks can be sent to multiple buses simultaneously, so they might have more than one destination set.

5   In the Busses section, click the B3:SFX Out button to select the SFX bus.

6   Click the tracks that contain diegetic audio: STAGE-L, STAGE-R, WALLA-L, WALLA-R, and PIANO to route them to B3o.

7 Click the B4: MUSIC Out bus to select the final bus for non-diegetic effects and music, and then click the two SFX tracks and the MUSIC tracks to assign them to B4o.

TIP To remove a track from a specific bus, click the bus from the Available Tracks section of the Bus Assign window, and then click the track. Tracks already assigned to the selected bus will be removed.

Lastly, you still won't hear anything from your timeline unless you route the buses themselves to the stereo bus you're using as the main output.

8 In the Busses section, click the B1:STEREO MAIN Out button, and then click the DIALOGUE, SFX, and MUSIC buttons in the Available Tracks section to route them to B1o.

NOTE You cannot route a bus to itself, so you'll be unable to select the STEREO MAIN track with the B1 STEREO MAIN Out bus also selected.

9 Once you have the buses correctly routed as in the preceding image, click Close.

Mixing with the Mixer

Wow! It might seem like a lot of hard work, but now everything will start to pay off since it's now easier to mix the audio for the scene just by using the controls for each of the buses.

1   In the mixer, unmute A10, A11, and A12.

> **TIP** Shift-clicking the Mute button in the timeline track controls for any muted track will unmute all currently muted tracks.

2   Play through the timeline from the beginning, listening to the various elements in the soundtrack.

What are your ears telling you? Most likely, you'll identify that the sound effects and music toward the end are far too loud!

3   Play the timeline again, and this time use the Music bus slider to bring the level of the music down by about 10 dB (or whatever level you think is suitable that you can hear the dialogue at the scene's climax).

4   Using the same technique, adjust the level of the SFX bus to reduce the sounds of the applause, atmos, and piano tracks together (if in doubt, try about -6 dB).

As you can see (and hear), using the buses has made it so much easier to mix the audio for the scene. What's more, if you need any greater degree of control (for instance, you might think that the radio dialogue track needs further changes), then you can always adjust the level of a specific track, or you can even go back into the clips themselves and further refine their individual levels if needed!

Now that you've seen the power of buses in DaVinci Resolve, mixing the audio in your timeline should be so much easier.

Using Buses for Alternative Mixes

A common requirement for many editors responsible for a film's audio is to set up the timeline for delivering different audio mixes. In the final exercise in this lesson, you'll create two additional buses for the SYNC SCENE FINAL MIX timeline: the first is for delivering an M&E (music and effects) version of the soundtrack, and the second will enable you to deliver a 5.1 surround-sound mix of the same soundtrack!

You'll start by creating two additional buses that you will use for each of the mixes.

1   Choose Fairlight > Bus Format to reopen the Bus Format window containing the timeline's current buses.

2   Click Add Bus twice to add two additional buses: Bus 5 and Bus 6.

3   Change the name of Bus 5 to **M&E**, the format to Stereo, and the color to Purple.

4   Change the name for Bus 6 to **SURROUND**, the format to 5.1, and the color to Violet.

5   Drag the M&E bus so that it's arranged below the STEREO MAIN bus.

6 Drag the SURROUND bus to below the M&E bus.

> **TIP** The arrangement of the buses in the Bus Format window reflects the order of the controls shown in the mixer, so this is a good practice to keep similar-use buses together.

7 Click OK to confirm your changes.

8 Expand the mixer so you can see all the controls.

The new buses appear in the mixer but, as before, nothing is being routed to them.

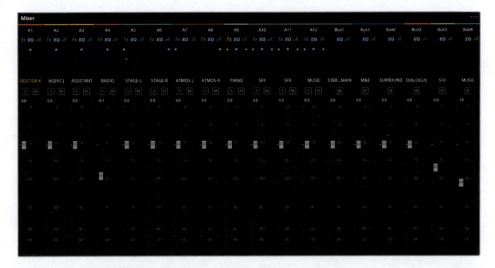

9   Choose Fairlight > Bus Assign to reopen the Bus Assign window.

10 In the Busses section, select the B5:M&E Out bus.

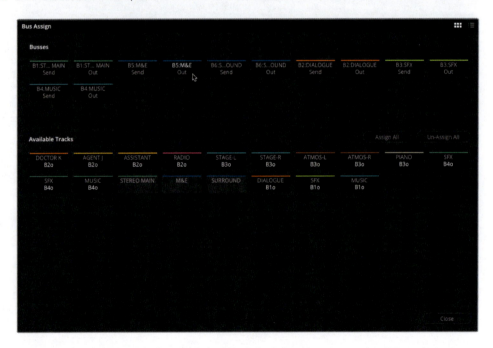

To create the M&E output for this bus, all you need to do is route the SFX and Music buses you created earlier, rather than the individual tracks.

11 In the Available Tracks section, select the SFX and MUSIC, so both buses are now being routed to the B1o and B5o buses.

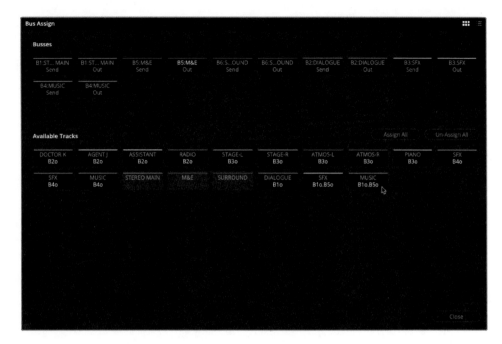

Next, you'll need to route the full mix to the SURROUND bus.

12 Click to select the B6:SURROUND Out bus from the Busses section.

13 From the Tracks section, click the DIALOGUE, SFX, and MUSIC tracks to route those buses to B6o.

14 Once you have the routing set as in the preceding image, click Close to close the Bus Assign window.

Monitoring the Different Buses

To listen to the different mixes from the same timeline, you just need to change the bus you're listening to.

1 Return your playhead to the start of the timeline and begin playing the scene.

2 From the Control Room pop-up menu above the mixer, select the M&E bus.

This is the soundtrack, minus the dialogue tracks.

3 Select the DIALOGUE bus from the Control Room pop-up menu.

You're now listening to *just* the dialogue bus for this scene.

4 Change the Control Room to the SURROUND bus.

You're now listening to the same mix play as a surround mix. Unfortunately, unless you have a surround system attached to your system and set up correctly for Resolve to use, you won't hear any change, but you can see the meters moving in the mixer. However, you'll need to make some further changes to how the audio is handled within the timeline since three channels in the 5.1 main are not currently being used.

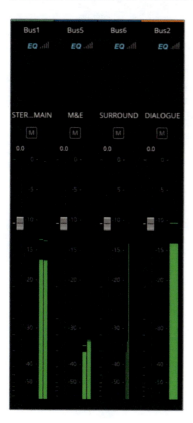

This is because the buses for the effects and music are both stereo.

5 Select Fairlight > Bus Format to reopen the Bus Format window.

6 Change the format for the SFX bus and the MUSIC buses to 5.1 and click OK.

By changing the buses from stereo to 5.1, it now means that any audio using the surround panners will have the audio correctly routed.

7 In the mixer, adjust the pan for the STAGE-L, STAGE-R, ATMOS-L and ATMOS-R tracks so they are midway between the front and rear speaker channels and then play the timeline again.

Now you can see that the two rear channels of the 5.1 main are being used (even if you can't actually *hear* them.

8 Drag the blue handle for the PIANO track to the bottom of the pan controls so that it will now only play out of the rear speakers.

For anyone listening in 5.1, it will now sound as though the piano player is behind them.

9 Double-click the A10 SFX track's pan controls.

The A10 pan controls open in a separate, larger window.

10 Drag the blue control into the center of the pan window so the Front/Back controls are set to C (for center).

11 In the Boom settings, click On to send audio to the subwoofer channel and set the level to -20.

12 Close the Audio Pan - SFX window.

The mixer's pan controls update to show the change you've just made.

13 Repeat the previous steps for the A11 and A12 tracks.

14 Play the timeline once more to see how the channels in the SURROUND bus are now being fully used.

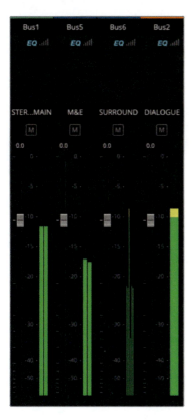

15   Return the Control Room pop-up menu to STEREO MAIN for the full stereo mix.

> **NOTE**   If you need to catch up before moving to the next lesson, select the Timelines bin and choose File > Import > Timeline, navigate to R18 Editors Guide/Lesson 08/Timelines and select SYNC SCENE FINAL MIX CATCHUP FINISHED.drt and click Open.

Congratulations! You just completed some high-end professional audio post-production from the comfort of your own computer workstation. Hopefully, this lesson has opened your eyes (and ears) to the wonderful, yet often underappreciated, world of audio post-production, as well as the sophisticated audio tools and configurations available as standard in DaVinci Resolve.

More information on working with audio in DaVinci Resolve is available in The Fairlight Audio Guide to DaVinci Resolve 18.

Lesson Review

1   True or false? Normalizing audio levels is a quick way of ensuring that all your audio is at the correct level.

2   True or false? You can change the speed of an audio clip in the edit page.

3   True or false? You can only mute or solo a track from the track header controls in the timeline.

4   Which window allows you to add and remove different buses?

   a)   Bus Assign

   b)   Bus Format

   c)   Audio Mixer

5   True or false? Fairlight FX must be applied in the Fairlight page.

Answers

1 False. Normalizing audio will adjust the level of a clip so that the peak is at the level specified.

2 True. To adjust the speed of any audio clip, you can use the Speed Change controls in the Audio tab in the Inspector.

3 False. You can also use the same controls in the mixer or the Tracks Index.

4 b) The Bus Format window allows you to add, change, and remove buses.

5 False. Fairlight FX are available in the Effects Library in the cut, edit, and Fairlight pages.

Delivering Projects

In Lesson 2, you used the Quick Export function to output a single video file suitable for uploading to a social media or video sharing site. The Quick Export window is very useful as a way of generating a file that you can use either as a final deliverable or simply as a way of showing the director or client your current progress on an edit with the fewest number of mouse clicks. However, if you want more control over your output files, as well as the ability to batch process multiple files, then the deliver page is the place to conduct your business.

This lesson will shed some light on how to customize your export options, as well as how to include subtitles and different audio tracks and buses in your final output.

Time

This lesson takes approximately 45 minutes to complete.

Goals

Preparing the Projects

For this lesson, you'll use three different projects that you have previously worked on to explore the various options that DaVinci Resolve provides for delivering projects in a variety of formats. You'll begin by importing all three projects into a separate folder in the Project Manager.

1 Launch DaVinci Resolve and, in the Project Manager, click Add Folder.

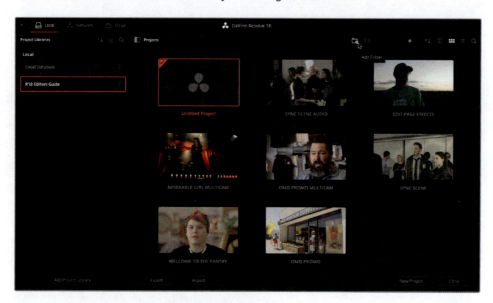

2 In the Create New Folder window, type **Delivery**, and then click Create.

A new folder called Delivery is created in the Project Manager.

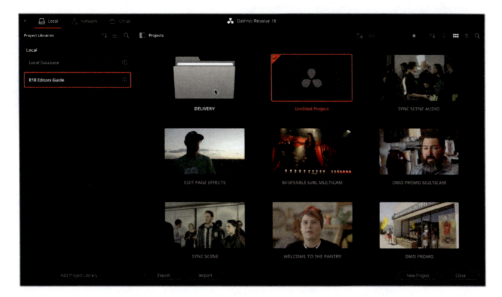

3   Double-click the Delivery folder to open it.

> **TIP** You can use the folder path in the top left of the Project Manager (Projects/DELIVERY) to navigate back out of this folder.

4   Open a new Finder window (macOS) or Explorer window (Windows) and navigate to R18 Editors Guide/Lesson 09.

This folder contains several files that you will use throughout this lesson.

5   Select the three DaVinci Resolve Project files (**OMO PROMO DELIVERY.drp**, **WTTP DELIVERY.drp**, and **SYNC SCENE DELIVERY.drp**) and drag them directly into the Project Manager window.

Each of the projects is imported into the Project Manager. You're now ready to continue with the lesson.

Reformatting a Timeline for Different Aspect Ratios

While most modern video cameras still shoot traditional 16:9 aspect video footage, that doesn't necessarily mean you are always required to deliver 16:9 footage. With many videos being watched on mobile devices, many social media content creators prefer an alternative aspect ratio, such as 1:1 (square formats such as those favored by Instagram) or 9:16 (vertical formats that are common on TikTok).

DaVinci Resolve allows you to customize your timeline settings for several different aspect ratios.

In the next exercise, you'll learn how easy it is to repurpose a timeline to fit a square aspect ratio.

1   From the Project Manager, open the **OMO PROMO DELIVERY** project and relink the media files.

2   If necessary, choose Workspace > Reset UI Layout.

3 From the Select Timeline dropdown menu, open the **OMO PROMO** timeline.

This is a finished version of the *Organ Mountain Outfitter's promo video* that you edited in Lessons 1 and 2.

4 Choose Timeline > Find Current Timeline in Media Pool.

5 With the timeline selected in the media pool, choose Edit > Duplicate Timeline and rename the duplicated timeline to **OMO PROMO SQUARE**.

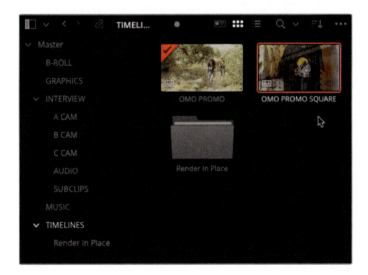

6 Double-click to open it in the timeline.

Next, you will need to adjust the settings of this timeline for the desired aspect ratio.

7 In the media pool, right-click the **OMO PROMO SQUARE** timeline and choose Timelines > Timeline Settings.

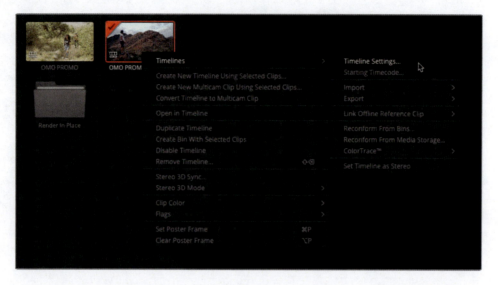

The Timeline Settings window appears. As you can see from the grayed-out Format, Monitor, and Output tabs, this timeline is using the timeline format for this project as set in the Project Settings.

8   Uncheck the Use Project Settings option.

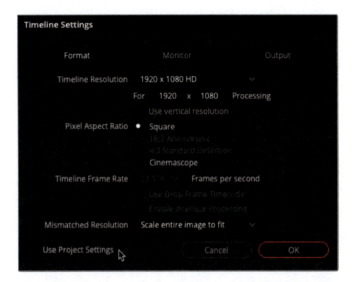

9   In the Timeline Resolution dropdown menu, choose the 1080 x 1080 HD Square option.

TIP   DaVinci Resolve contains presets for common timeline format resolutions, but you're free to choose a custom timeline resolution by choosing Custom from the preset menu and specifying the horizontal and vertical resolutions in the For Processing boxes. For vertical resolutions, simply check the "Use vertical resolution" checkbox to invert the current horizontal and vertical values.

10 Click OK to update the timeline settings for this timeline.

The timeline viewer shows the updated resolution. However, it has simply *letterboxed* the 16:9 footage into the square aspect ratio. The client does not want this "empty space" above and below the video, preferring that the video completely fill the square aspect ratio.

This is due to the fact that Resolve is resolution independent—that is, by default, it doesn't matter what resolution the source video is; it will always be formatted to fit the current timeline resolution. However, you can always override this in the timeline settings.

11 Right-click the **OMO PROMO SQUARE** timeline in the media pool again and choose Timeline > Timeline Settings to reopen the Timeline Settings window.

12 Change the Mismatched Resolution menu to "Scale full frame with crop" and click OK.

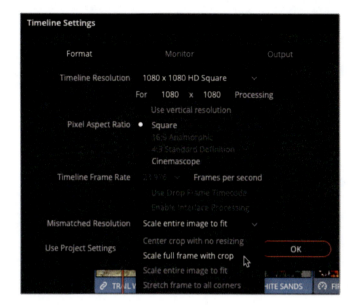

Now all the clips in this timeline correctly fill the 1:1 aspect ratio. However, unless the shots were specifically filmed with repurposing to a square aspect ratio in mind, it's likely that not all of them will be framed correctly.

Reframing Shots

Unless the director of photography (DoP) had a clear understanding that all or part of the final film would need to be displayed in a 1:1 aspect ratio, it's very likely that you will need to adjust the framing of some of these shots. Thankfully, you always have access to the full picture information, even though it appears currently cropped.

1   In the square timeline, place the playhead over the second interview clip.

Because the clip was originally shot off-centered, half of Chris's head is now disconcertingly missing off the side of the frame.

2   Press Shift-` (grave accent) or choose View > Viewer Overlay > Transform to display the Transform overlays.

3 Hold Shift and drag the shot of Chris in the timeline viewer to center him in the square frame.

> **NOTE** Alternatively, you can use the Position X parameter in the Transform section of the Inspector if you prefer.

4 Play through the next two clips on V2 and adjust the horizontal framing, if required.

While you can often get away with simply adjusting a clip's Position X value, other shots might be a bit more problematic.

5 Play the next clip on V2, **WHITE SANDS**.

This is one such problematic shot, since the DoP framed this shot for the 16:9 aspect ratio of HD video, not considering the possible need to repurpose this shot for a square aspect ratio, leading to the slightly awkward framing of the four friends.

6   Place the timeline playhead on the first frame of the **WHITE SANDS** clip and open the Inspector.

7   Adjust the Position X value to about 150.000 to frame the first three friends to the left.

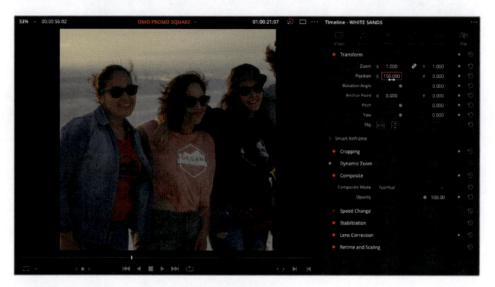

8   Click the Add Keyframe button for the Position values.

9   In the timeline, disable the Auto Track Selector control for V1.

10 Press ' (apostrophe) to jump to the last frame of the clip on V2.

> **NOTE** Leaving the Auto Track Selector enabled for V1 would mean that Resolve would have used the duration of the clip on V1 when you moved to the last frame.

11 Change the Position X value to about -20.000 to frame the three friends to the right.

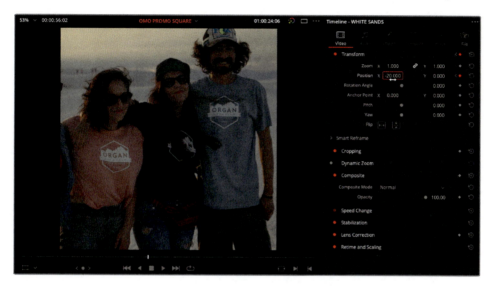

12 Play the shot to review the "pan" you have just created.

For a more "natural" movement to the pan, you can add a touch of smoothing to the keyframes' acceleration.

13 Make sure your playhead is over the WHITE SANDS clip and press [(left square bracket) to jump back to the first keyframe.

14 In the Inspector, right-click the red keyframe button next to the Position values and choose Ease Out to apply a default Bézier curve to the right of this keyframe.

15 Press] (right square bracket) to jump to the next keyframe, right-click the red keyframe button for the Position values, and choose Ease In to add a default Bézier curve to the left of this keyframe.

NOTE Ease Out is available only if there is another keyframe later in the clip and Ease In is available only if there is another keyframe earlier in the clip. If there are other keyframes before and after the current keyframe, you will also see the option for Ease Both. Select the clip and press Shift-C to reveal the clip's keyframe curve in the timeline to manually add or adjust the Bézier handles as required.

By easing the keyframes, you apply a slightly smoother change to the animation.

Smart Reframe (Studio Only)

While the ability to manually add keyframes and adjust the framing of a clip is no doubt useful, if you must do this for a large number of shots it can quickly slow down the process of preparing the video for export. If you have several shots to reframe in this manner, using the Smart Reframe feature in DaVinci Resolve 18 can be a huge timesaver.

1   In the **OMO PROMO SQUARE** timeline, place the playhead over the **FIRE DANCER** clip.

This is another shot that could do with a small amount of reframing.

2   In the Inspector, click the disclosure arrow for the Smart Reframe controls and then click the Reframe button.

After a short analysis, the shot reframes itself, centered on the dancer!

Smart Reframe uses the DaVinci Resolve Neural Engine to analyze the clip and determine the best way to reframe for any given shot.

3   Select the **FIRE DANCER** clip in the timeline, choose Clip > Show Keyframe Editor, and click the disclosure arrow for the Transform settings.

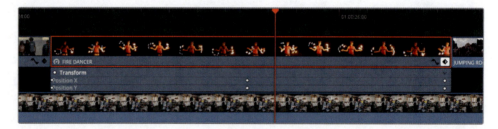

> **TIP** If you can't see the keyframes in the Keyframe Editor, zoom in on the clip in the timeline.

The Smart Reframe function has automatically added keyframes to the Position X and Y parameters in order to keep the clip framed.

You can also choose an area of interest for the Smart Reframe function to focus on.

4   Press Shift-C to hide the Keyframe Editor.

5   In the timeline, select the **COMPUTER DESIGN** clip.

6 In the Smart Reframe controls in the Inspector, change the Object of Interest menu to Reference Point.

The Reference Point area is displayed in the timeline viewer as a box.

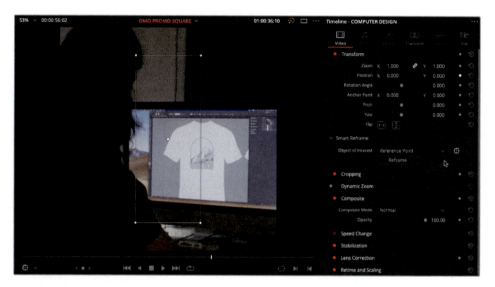

7 Resize the Reference Point box in the timeline viewer so that it frames the T-shirt design on the screen.

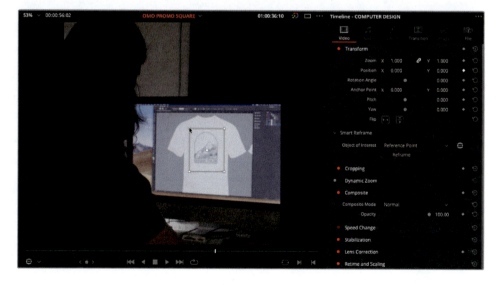

8   Click the Reframe button in the Smart Reframe controls.

The Smart Reframe function now keeps the framing on the design.

As you can see, the Smart Reframe function in DaVinci Resolve Studio can save time by quickly reframing shots for you.

> **TIP** You can select multiple timeline clips and use the Smart Reframe's Auto feature on all of them together. Each clip will be analyzed separately for the best results.

Customizing Deliver Page Presets

Instead of using the Quick Export feature to output this video, you'll use the options in the deliver page, which provide more customization and control over the final exported file.

1   Click the Deliver page button.

The square timeline opens in the deliver page, where you can assign much more specific settings to the file you want to output.

Render Presets—Quick
access to saved presets

Render Settings—Used to customize Video,
Audio and File settings for the current preset

Render Queue—List of any
jobs ready to be rendered

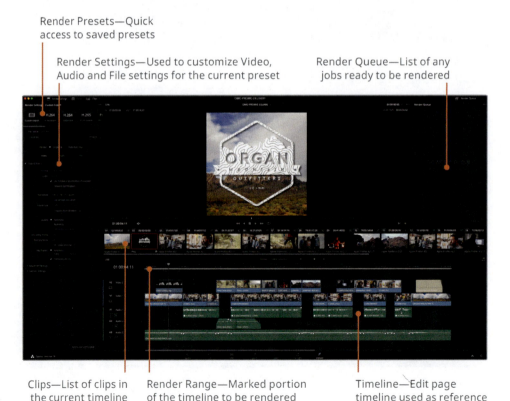

Clips—List of clips in
the current timeline

Render Range—Marked portion
of the timeline to be rendered

Timeline—Edit page
timeline used as reference

2   To export your square video for Instagram or other social media sites that utilize square-aspect videos, select the H.264 Master preset.

3   In the Filename field, type **OMO PROMO SQUARE**.

4   For the Location, click the Browse button, and in the File Destination window that appears, navigate to R18 Editors Guide/Output Folder, and click Save.

5   In the Video tab, check that the resolution is set to 1080 x 1080, and the frame rate is 23.976.

6 Click the Audio tab and verify that the codec is AAC, the data rate is 320 Kb/s, and the output track is Bus 1 (which is the only bus for this timeline).

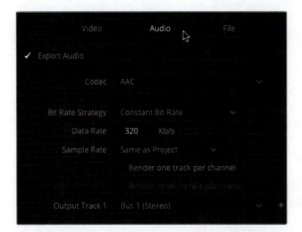

7 Click Add to Render Queue at the bottom of the Render Settings panel.

The job is added to the Render Queue as Job 1. From here, you could choose to render out the job, but there are more jobs to add to the queue before you start the rendering.

Understanding Data Levels

In the Advanced Settings, Data Levels specifies the data range of an image based on its source. The default Auto setting renders media at a data level appropriate for the selected codec. Video refers to YCbCr formats that constrain to pixel data values between 64–940 on a 10-bit system in formats using a Rec.709 video standard. Full expands the range to the film standard of 4–1024 values, which is utilized in digital film formats such as DPX. In general, the best choice is to leave this setting unchanged and let DaVinci Resolve choose the data level automatically. However, if you find that your final video looks substantially darker or lighter than it appears on your calibrated monitor, it's possible that the data levels are being incorrectly distributed. In that rare case, you might want to manually set the Data Levels correctly for your intended distribution.

Exporting AAF for Pro Tools

Despite Fairlight having all the tools necessary to mix and master your audio, another task required of many editors is to send the audio to be mixed in a Pro Tools system. In this exercise, you'll look at using the Pro Tools export preset for doing just that.

1   Press Shift-1 to open the Project Manager.

2   Double-click the WTTP DELIVERY project. If prompted, save the current project to prevent the loss of your changes.

3   Switch to the edit page and use the Relink Media button to relink the files.

   This project is a version of the *Welcome to the Pantry* promo you edited in Lesson 3. You might remember that you'd didn't spend too much time finessing the audio for this project. So now you will export it in a format suitable for a Pro Tools system, where the audio will be mixed for you.

4   Click the deliver page again, and from the Render Settings panel, scroll along and select the Pro Tools preset.

> **NOTE** Resolve uses the AAF (Advanced Authoring Format) to output timeline information for Pro Tools systems.

Presets like Pro Tools don't allow you to set a filename because you're not rendering a single file. Instead, you're rendering out multiple files for each of the clips in the timeline, which in the case of the Pro Tools preset, is all the audio clips.

5   Click Browse and navigate to the R18 Editors Guide/Output Folder. Create a new folder here called **Pro Tools AAF** and click Open.

6   In the Audio tab, ensure that the codec is set to Linear PCM. This will create a series of uncompressed .wav files for the audio clips in your timeline, as preferred by many Pro Tools users.

7   Choose to add 12 frame handles to the media you will render.

Although you're rendering out the audio clips in this timeline for Pro Tools, you should also provide a video reference file for the Pro Tools operator.

8   Click the Video tab. Change the Format to QuickTime, and then ensure that the codec is H.264, and the resolution is 1280 x 720 HD 720P (the resolution of this timeline).

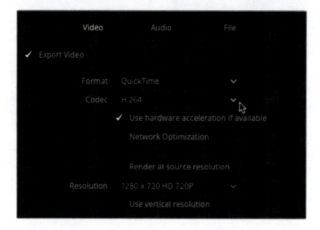

It would also be useful to have Burned-in Timecode (BiTC) on the video reference clip.

9   Choose Workspace > Data Burn-In.

10 In the Data Burn-In window, select Record Timecode to display the timeline's timecode. Increase the size to about 85 and the background opacity to 80, and then close the Data Burn-In window.

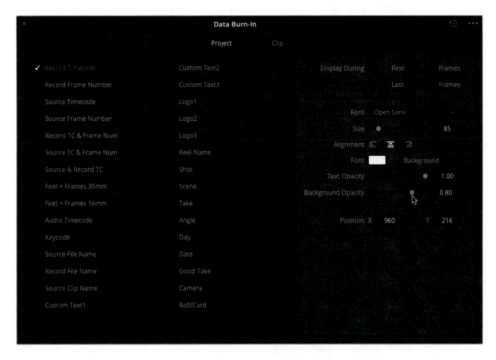

The Data Burn In will appear in the viewer as you make the changes.

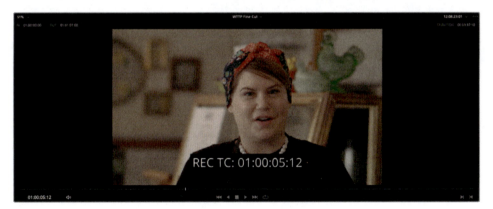

11 Click Add to Render Queue to add this job to the Render Queue window.

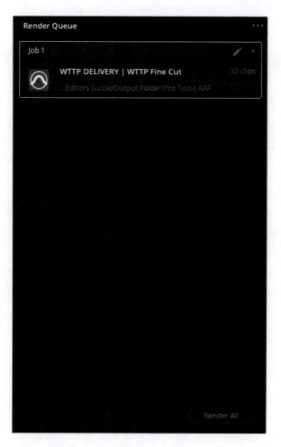

That's two jobs now queued and ready to go. There's just one more job to add to the Render Queue, and then you can let Resolve render all the jobs together.

Adding Subtitles

DaVinci Resolve allows you to add subtitles to your timelines in several ways. You can manually create all your subtitles for your project, or you can import a supported subtitle file.

1 Press Shift-1 to open the Project Manager and double-click the SYNC SCENE DELIVERY project, saving the changes to the current open project if prompted.

2 Press Shift-4 to jump to the edit page and relink the media for this project.

This project contains a finished version of the Sync scene that you've worked on in previous lessons. Take a moment to reacquaint yourself with the scene and how the audio is mixed. All that's left for you to do is add subtitles before outputting the final files for delivery.

3 Open the Effects Library.

4 In the Effects Library, in the Titles group, locate the Subtitle title.

5 Drag the Subtitle to the timeline in the space above the video tracks and snap it to the beginning of the timeline.

A new track appears in the timeline labeled ST1 Subtitle 1, and the subtitle text appears in the timeline viewer.

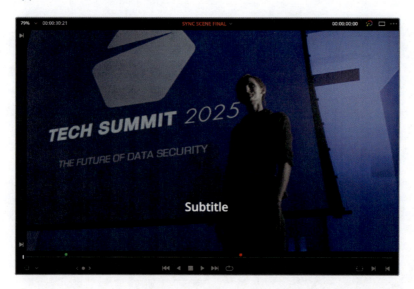

> **TIP** You can show and hide existing subtitle tracks in the Timeline View Options menu.

6   In the timeline, select the subtitle, and then open the Inspector.

The Inspector includes the controls for the individual subtitle captions and for the entire subtitle track.

7   In the Caption field, highlight the word "Subtitle," and type **[applause]**.

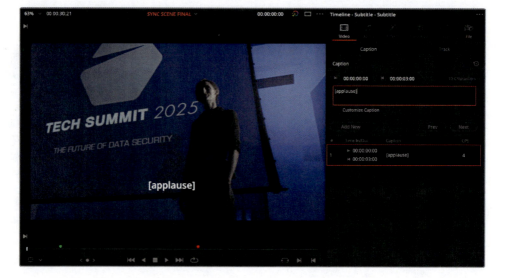

8 Place the timeline playhead at the start of the next timeline clip on V1 and, in the Inspector, click the Add New button.

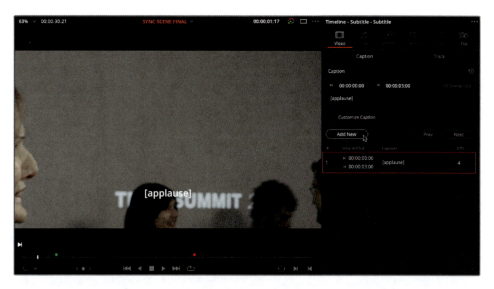

A new subtitle is overwritten onto the subtitle track at the current playhead position.

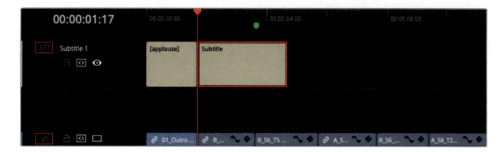

9 Select the new subtitle and, in the Inspector, type the dialogue from the doctor for the next clip: **Oh, thank you. I'm so glad you really liked it**.

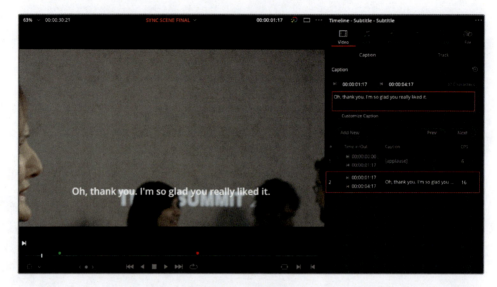

10 Trim the end of the second subtitle to the end of the second clip on V1.

You can edit and trim Subtitle generators just like any other clip on your timeline.

Importing Subtitle Files

You can continue working through this timeline, adding subtitles for the different lines of dialogue. However, it's usually much more efficient and accurate to have someone transcribe the dialogue for you and create a subtitle file that you can import directly.

> **NOTE** DaVinci Resolve supports the importing and exporting of .srt, .vtt, .ttml (IMSC1), and .dfxp subtitle files.

1   In the media pool, select the Subtitles bin and choose File > Import > Subtitle.

2   Navigate to R18 Editors Guide/Lesson 10/Subtitles and select the file
    SYNC SUBTITLES US.srt. Click Open.

    The .srt file is added to the selected bin as a subtitle clip.

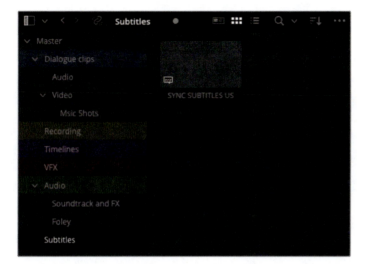

3   Select the subtitle file SYNC SUBTITLES US and drag it into the timeline, so it snaps to
    the green timeline marker.

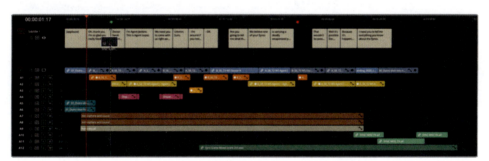

All the subtitles in the .srt file are added to the Subtitle 1 track in the timeline.

Adjusting Subtitles

Subtitles clips behave just like any clip in the Resolve timeline, so you can easily adjust their timings as necessary.

1   Move the timeline playhead to the red marker, where the doctor says, "That wouldn't be possible." Play the timeline from this point to review the dialogue and subtitles.

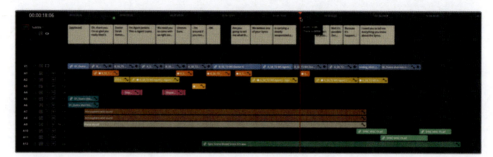

The subtitles appear onscreen too late for the doctor's line of dialogue. You'll need to adjust the subtitle timing to sync it correctly.

2   Press T to enter Trim Edit mode and slide the subtitle to the left by about a second, until it snaps to the beginning of the doctor's audio clip on A1.

3   When you've finished sliding the subtitle into the correct position, press / (forward slash) to review the change.

4   Press A to return to Selection mode.

You can also trim and roll trim each of the subtitles using the Selection mode, as well as cut each subtitle into shorter clips using the Blade Edit mode and keyboard shortcuts.

Styling Subtitles

Just as with any other title in Resolve, subtitles have many parameters that allow you to adjust the style and position of your subtitle text. One common style applied to subtitles is a semi-transparent box behind the text to help it stand out against video with a similar brightness.

1   Move your playhead to the start of the last subtitle in your timeline.

    The end of the line appears a little obscured against this final shot and isn't easy to read.

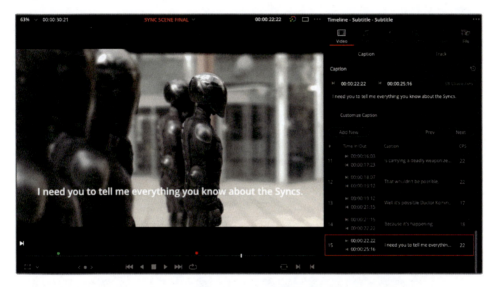

2   In the timeline, select the subtitle, and in the Inspector, click the Track tab.

3   Scroll down to the Background options and enable the Background settings.

A semi-transparent box appears around the subtitle's text, helping it stand out from the similarly light video background.

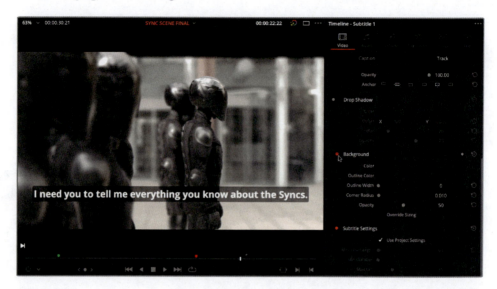

Because these changes have been made in the Track tab, all the subtitles in this track are updated with the change. This behavior is particularly useful when you need to modify style settings for all the subtitles in a track.

You can, however, override those track-wide settings for any individual subtitle when you need to adjust the color, font, or position of one or more subtitles, but not all.

4   In the timeline, move the playhead to the first subtitle.

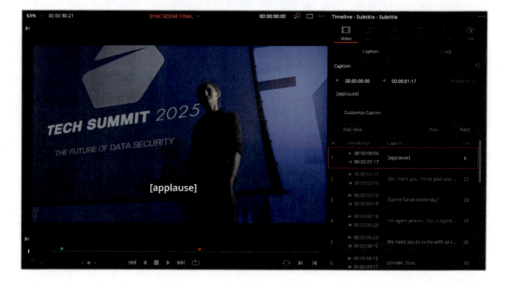

This subtitle indicates a sound effect rather than detailing spoken dialogue. As such, the director would like to use a different style to differentiate it from captions for dialogue.

5   In the timeline, select the subtitle, and in the Inspector, select the Caption tab.

6   Check the Customize Caption box to reveal an additional set of controls for this caption in the Caption tab.

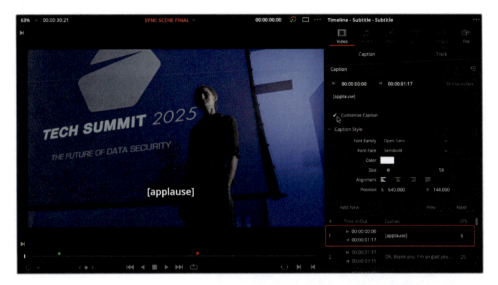

7   Change the Font Face setting to Semibold Italic to distinguish this caption from the other captions.

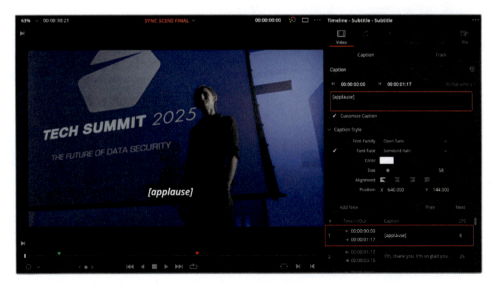

The subtitle updates to reflect the change in style, but the other subtitles in the track do not change.

You can add multiple subtitle tracks, which is particularly useful when you need to supply subtitles in more than one language.

8   In the edit page, right-click any one of the timeline track headers and choose Add Subtitle Track to add an additional subtitle track to the timeline.

9   With the Subtitles bin selected in the media pool, choose File > Import > Subtitle.

10 From the R18 Editors Guide/Lesson 10/Subtitles, select the **SYNC SUBTITLES FR.ttml** file and click Open to add this subtitle to the media pool.

11 Edit this new subtitle file into the empty subtitle track you just created, starting at the beginning of the timeline.

This subtitle file includes French subtitles. You can rename subtitle tracks to reflect the language used, making it easier to identify the different tracks.

12 Click the Subtitle 1 track name, and type **en_US** to identify this subtitle as English for a US audience.

13 Click the Subtitle 2 track name, and type **fr_FR** to identify this subtitle as French for a French audience.

> **TIP** To choose the subtitle track visible in the timeline viewer, in the head of the track you wish to view, click the eye icon. You can display only one subtitle track at a time.

Using Subtitle Regions

Instead of customizing individual subtitles, you can set subtitle regions. Each subtitle track can have up to four regions, with each region having its own style. Moving a subtitle from one region to another instantly updates the subtitle to that region's settings.

To create a new region, right-click anywhere along the subtitle track and choose Add Subtitle Region. Move subtitles into the region by dragging them up or down in the subtitle track.

Subtitle regions are supported only by .ttml and .dfxp subtitle files. Make sure those file types are supported by your destination of choice before submitting your video file with subtitles.

Exporting with Subtitles

When it comes to delivering subtitles with your finished program, several options are open to you. Depending on your delivery format, you can include subtitles as burned-in graphics, embedded text in a supported media file, or as a separate file.

In this next exercise, you'll output a file for delivery, together with the necessary separate subtitle files.

1   Click the Deliver page button and choose the Custom preset.

2   Set the render to "Single clip" to export a single movie file.

3   In the File Name field, type **SYNC SCENE MASTER**.

4   Click the location Browse button and navigate to the R18 Editors Guide/Lesson 09/ Output folder. Create a new folder called **Sync Scene** and click OK.

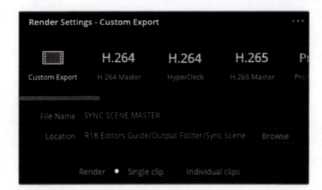

5   Back in the Render Settings panel, set the Format dropdown menu to MXF OP1A, the Codec to DNxHD, and the Type to 720p 100/85/55/45 8-bit.

6   At the bottom of the Video tab, expand the option for Subtitle Settings, and choose Export Subtitle.

7   Set the Format option to "As a separate file."

If you chose the "Burn into video" option, Resolve would burn the currently active subtitles (with their styles) into the final rendered video file, so the subtitles would be permanently included as part of the video content. Choosing "As embedded captions" will output the currently active subtitle track as an embedded metadata layer within those media formats that support it. Currently, DaVinci Resolve supports CEA608 and text captions within MXF OP1A and QuickTime containers.

8   In the Export As dropdown menu, choose "SRT" and select both the "en_US" and "fr_FR" subtitle tracks to include them in the export.

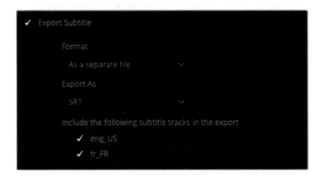

This takes care of the video and file settings. Next, you'll choose which audio tracks to output as part of this master file.

Exporting Audio Tracks

You will remember from the previous chapter that this scene contains three different mixes: a stereo mix, M&E (music and effects), and a 5.1 mix, together with three buses for dialogue, effects, and music only. In the Render Settings, you can choose to output different combinations of audio tracks and buses. Doing so makes it incredibly easy to export a final movie with stereo, surround, and an M&E mix for international dubbing, or just the 5.1 mix.

1   In the Render Settings, click the Audio tab.

2   Leave the audio codec set to Linear PCM (this is also referred to as *uncompressed, or lossless*, audio) and leave the sample rate and bit depth settings as they are.

> **TIP** Project sample rates default to 48 kHz but can be set in Project Settings > Fairlight.

3   From the Output Track 1 dropdown menu, ensure that the STEREO MAIN (Stereo) bus is selected.

4   Click the + (plus) button to the right of the Output Track 1 menu to add a second output track.

5   From the Output Track 2 menu, choose the M&E (Stereo) bus to add this bus as the second output track.

This will export the MXF file with two stereo tracks: the first for the full stereo mix; the second for the M&E mix, which does not contain the dialogue audio of the soundtrack—just the music and effects elements. It is the second track that can be used when reversioning the film for a different language.

6   Click Add to Render Queue.

If you need to use the same settings for more than one export, consider saving the render settings as a preset.

7   Click the Render Settings options menu and choose Save As New Preset.

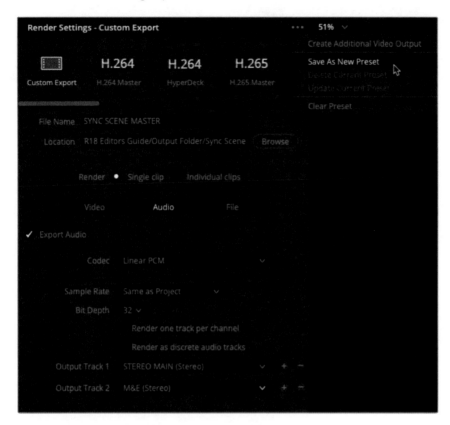

8   Name your preset **MXF DNxHD 720P** and click OK.

The new preset appears at the top of the Render Settings panel.

Changing and Rendering Jobs from Multiple Projects

The Render Queue can show jobs from the current project or from all projects in your database. If you split longer projects into reels, or you are working on different projects for the same client, you might need to access all the jobs in the queue instead of waiting for one batch to render before outputting other projects.

1   In the Render Queue options menu, choose Show All Projects.

Any jobs added to the Render Queue in any project in the current project library are displayed for you to select and render.

2   In the Render Queue options menu, choose Show Job Details.

The specific settings for each job are displayed, including resolutions, codec, and frame rate.

Even after you add jobs to the Render Queue, you can update their settings or remove them from the queue entirely.

3 In the Render Queue, click the pencil icon for Job 1. If prompted, save the current project.

The **OMO PROMO DELIVERY** project reopens automatically, and the current render settings for the selected job become available.

Now you can make any changes to the job before rendering out the final file.

4 In the Video tab, change the Format to MP4 instead of QuickTime.

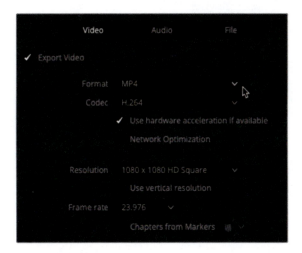

5 Click Update Job at the bottom of the Render Settings panel.

The change updates the original job settings with the new setting as reflected in the displayed job details in the Render Queue. In this case, you can see the file extension change from .mov to .mp4.

> **TIP** To delete a job, click the "X" in the upper-right corner of the job in the Render Queue.

6   Finally, click in an empty area of the Render Queue panel to deselect Job 1, and then
    click the Render All button to create the output files.

DaVinci Resolve renders each of the jobs in the Render Queue.

Once all the files have been rendered, you can choose to open the Output Folder on your system and verify the files you've created.

7   Right-click any of the completed jobs and choose Reveal in Finder (macOS) or Open File Location (Windows).

> **NOTE** While you can often open and view many files to check their integrity (such as MP4 files, for example), the only true test of the success of creating the file(s) (such as with the AAF for Pro Tools) is to open them back into Resolve.

Utilizing the correct render settings is vital to delivering an aesthetically correct and technically functional video project. Understanding these settings has even greater benefits. It elevates your skill set as an editor and imbues confidence that your projects are delivered at their optimal quality and adhere to industry standards.

Media Managing Timelines

A final step in the delivery options for your projects is to manage the source media files for your projects for easy archiving.

1   If necessary, press Shift-1 to open the Project Manager and double-click SYNC SCENE DELIVERY to reopen the Sync Scene project.

> **TIP** You can enable Dynamic Project Switching by right-clicking in an empty area of the Project Manager to allow you to quickly switch between projects without having to close the current project.

2   Choose File > Media Management to open the Media Management window.

Media Management allows you to copy or transcode the media for the entire project, specific timelines, or specified clips.

3   Select the Timelines tab.

4   Next to the Destination field, click Browse.

5   Navigate to the R18 Editors Guide/Lesson 10/Output Folder and create a new folder called **Archive**. Click Open.

6   In the Media Management window, select the SYNC SCENE FINAL timeline.

7   In the Copy tab, make sure the "Used media and trim keeping 24 frame handles" option is selected.

This option means the media files being copied will only be the parts of those files actually used in the selected timeline, plus an additional 24 frames before the In point and after the Out point, just in case a small tweak is ever needed.

8   Uncheck the "Use project name subfolder" option.

At the bottom of the Media Management window, the current size indicator displays the total storage size of all the media currently in this project (including unused clips). The new size indicator shows the amount of storage the trimmed media will use. In this case, you can see that the copied media will only require about a third of the storage of the full project.

9 Click Start to begin the media management process.

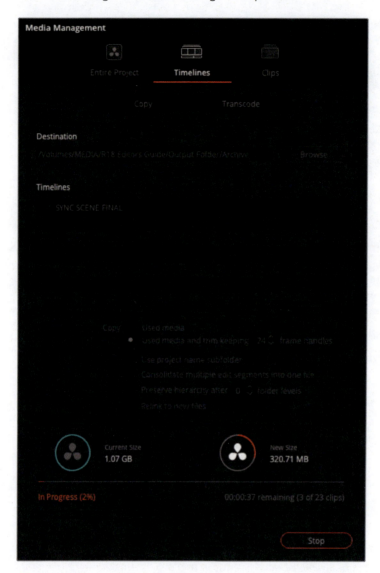

10 Once complete, open the Archive folder to find the copied media and a .drt file.

This is the folder that you can now archive separately. To restore the edit from this archived material, simply create a new project and choose File > Import > Timeline and import the .drt file. The imported timeline will automatically relink to the trimmed media.

Lesson Review

1   True or false? You must use the deliver page in order to export a video file from your project.

2   Which function uses the DaVinci Resolve Neural Engine to automatically add keyframes to keep the subject of a clip in frame when reformatting a timeline to a different aspect ratio?

    a) Auto Conform

    b) Smart Reframe

    c) Smart Conform

3   What format is commonly used when sending audio to be mixed on a Pro Tools system?

    a) AAF

    b) XML

    c) EDL

4   Which window enables you to add a Burned-in Timecode (BiTC) to your exported video?

    a) Timecode Window

    b) Data Burn-In

    c) Text+

5   True or false? All subtitles must be imported from a supported .srt file.

Answers

1   False. You can use the Quick Export option, where available.

2   b) Smart Reframe.

3   a) AAF (Advanced Authoring Format)

4   b) Data Burn-In

5   False. Subtitles can be manually created inside DaVinci Resolve using the Subtitle title in the Effects Library, or imported from a .srt, .vtt, .ttml, or .dfxp file.

Index

Made in the USA
Monee, IL
17 March 2024

55209511R10297